Deterministic Mathematical Models in Population Ecology

MONOGRAPHS AND TEXTBOOKS IN PURE AND APPLIED MATHEMATICS

1. *K. Yano,* Integral Formulas in Riemannian Geometry (1970)
2. *S. Kobayashi,* Hyperbolic Manifolds and Holomorphic Mappings (1970)
3. *V. S. Vladimirov,* Equations of Mathematical Physics (A. Jeffrey, editor; A. Littlewood, translator) (1970)
4. *B. N. Pshenichnyi,* Necessary Conditions for an Extremum (L. Neustadt, translation editor; K. Makowski, translator) (1971)
5. *L. Narici, E. Beckenstein, and G. Bachman,* Functional Analysis and Valuation Theory (1971)
6. *D. S. Passman,* Infinite Group Rings (1971)
7. *L. Dornhoff,* Group Representation Theory (in two parts). Part A: Ordinary Representation Theory. Part B: Modular Representation Theory (1971, 1972)
8. *W. Boothby and G. L. Weiss (eds.),* Symmetric Spaces: Short Courses Presented at Washington University (1972)
9. *Y. Matsushima,* Differentiable Manifolds (E. T. Kobayashi, translator) (1972)
10. *L. E. Ward, Jr.*, Topology: An Outline for a First Course (1972) *(out of print)*
11. *A. Babakhanian,* Cohomological Methods in Group Theory (1972)
12. *R. Gilmer,* Multiplicative Ideal Theory (1972)
13. *J. Yeh,* Stochastic Processes and the Wiener Integral (1973) *(out of print)*
14. *J. Barros-Neto,* Introduction to the Theory of Distributions (1973) *(out of print)*
15. *R. Larsen,* Functional Analysis: An Introduction (1973)
16. *K. Yano and S. Ishihara,* Tangent and Cotangent Bundles: Differential Geometry (1973)
17. *C. Procesi,* Rings with Polynomial Identities (1973)
18. *R. Hermann,* Geometry, Physics, and Systems (1973)
19. *N. R. Wallach,* Harmonic Analysis on Homogeneous Spaces (1973)
20. *J. Dieudonné,* Introduction to the Theory of Formal Groups (1973)
21. *I. Vaisman,* Cohomology and Differential Forms (1973)
22. *B.-Y. Chen,* Geometry of Submanifolds (1973)
23. *M. Marcus,* Finite Dimensional Multilinear Algebra (in two parts) (1973, 1975)
24. *R. Larsen,* Banach Algebras: An Introduction (1973)
25. *R. O. Kujala and A. L. Vitter (eds.),* Value Distribution Theory: Part A; Part B. Deficit and Bezout Estimates by Wilhelm Stoll (1973)
26. *K. B. Stolarsky,* Algebraic Numbers and Diophantine Approximation (1974)
27. *A. R. Magid,* The Separable Galois Theory of Commutative Rings (1974)
28. *B. R. McDonald,* Finite Rings with Identity (1974)
29. *J. Satake,* Linear Algebra (S. Koh, T. Akiba, and S. Ihara, translators) (1975)

30. *J. S. Golan,* Localization of Noncommutative Rings (1975)
31. *G. Klambauer,* Mathematical Analysis (1975)
32. *M. K. Agoston,* Algebraic Topology: A First Course (1976)
33. *K. R. Goodearl,* Ring Theory: Nonsingular Rings and Modules (1976)
34. *L. E. Mansfield,* Linear Algebra with Geometric Applications (1976)
35. *N. J. Pullman,* Matrix Theory and Its Applications: Selected Topics (1976)
36. *B. R. McDonald,* Geometric Algebra Over Local Rings (1976)
37. *C. W. Groetsch,* Generalized Inverses of Linear Operators: Representation and Approximation (1977)
38. *J. E. Kuczkowski and J. L. Gersting,* Abstract Algebra: A First Look (1977)
39. *C. O. Christenson and W. L. Voxman,* Aspects of Topology (1977)
40. *M. Nagata,* Field Theory (1977)
41. *R. L. Long,* Algebraic Number Theory (1977)
42. *W. F. Pfeffer,* Integrals and Measures (1977)
43. *R. L. Wheeden and A. Zygmund,* Measure and Integral: An Introduction to Real Analysis (1977)
44. *J. H. Curtiss,* Introduction to Functions of a Complex Variable (1978)
45. *K. Hrbacek and T. Jech,* Introduction to Set Theory (1978)
46. *W. S. Massey,* Homology and Cohomology Theory (1978)
47. *M. Marcus,* Introduction to Modern Algebra (1978)
48. *E. C. Young,* Vector and Tensor Analysis (1978)
49. *S. B. Nadler, Jr.,* Hyperspaces of Sets (1978)
50. *S. K. Sehgal,* Topics in Group Rings (1978)
51. *A. C. M. van Rooij,* Non-Archimedean Functional Analysis (1978)
52. *L. Corwin and R. Szczarba,* Calculus in Vector Spaces (1979)
53. *C. Sadosky,* Interpolation of Operators and Singular Integrals: An Introduction to Harmonic Analysis (1979)
54. *J. Cronin,* Differential Equations: Introduction and Quantitative Theory (1980)
55. *C. W. Groetsch,* Elements of Applicable Functional Analysis (1980)
56. *I. Vaisman,* Foundations of Three-Dimensional Euclidean Geometry (1980)
57. *H. I. Freedman,* Deterministic Mathematical Models in Population Ecology (1980)

Other Volumes in Preparation

Deterministic Mathematical Models in Population Ecology

H. I. FREEDMAN
Department of Mathematics
University of Alberta
Edmonton, Canada

MARCEL DEKKER, INC. New York and Basel

Library of Congress Cataloging in Publication Data

Freedman, Herbert I
Deterministic mathematical models in population ecology.

(Pure and applied mathematics ; v. 57)
Bibliography: p.
Includes index.
1. Population biology--Mathematical models.
2. Ecology--Mathematical models. I. Title.
QH352.F73 574.5'248 80-19946
ISBN 0-8247-6653-9

MARCEL DEKKER, INC.
270 Madison Avenue, New York, New York 10016

Current printing (last digit):
10 9 8 7 6 5 4 3 2 1

PRINTED IN THE UNITED STATES OF AMERICA

To Donna, Dena,
Joseph, Tobey, and Daniel,
my life and inspiration

PREFACE

Mathematical ecology is a subject which lately seems to have interested many mathematicians and ecologists. The broad area of mathematical ecology can be divided into two broad subareas, statics and dynamics. The excellent book by E. C. Pielou deals mainly with the statics aspect, while the excellent books by R. M. May and J. Maynard Smith deal with the dynamics aspect from an ecologist's point of view. It is the intention of this book to deal with certain aspects of the dynamics aspect of mathematical ecology from a mathematician's point of view.

In a book of this size, one must always be concerned with which topics to include and which topics not to include. Some topics are not included because of the lack of available results. Other topics of interest are excluded because of lack of space or because of complexity. Basically, the models included are those which are given by autonomous ordinary differential or difference equations and are included because of my main interests. However, references are given for most of the relevant models known to the author, be they deterministic or stochastic.

The book is in three main sections, the introduction, predator-prey considerations, and competition and cooperation considerations. Each section contains two or more chapters. In each chapter there are references to many relevant models known to the author, as well as exercises. The exercises may contain some standard problems, some difficult problems, as well as some open (research) problems.

It is expected that the reader will have some background in analysis, including ordinary differential equations. Some of the more pertinent topics in these areas, e.g., perturbation theory and stability theory, are included in the appendix.

It is impossible to thank everyone who was involved in the writing of this book. However, three people must be singled out. For their encouragements and their constructive comments, I am forever grateful to P. E. Waltman of the University of Iowa and S. B. Hsu formerly of the University of Utah. I also wish to thank R. P. McGehee of the University of Minnesota for many inspirational conversations. Special thanks are also due to Mrs. G. Smith for typing the first draft of the manuscript and to Mrs. V. Spak for a masterful job in typing the final copy.

The book was begun while I was on sabbatical leave at the University of Minnesota and was completed at the University of Alberta. Special thanks are due to the mathematics departments of these two institutions for their support and help.

I also wish to acknowledge the Government of Canada which through its granting agency, the National Science and Engineering Research Council partially supported the research and preparation of this book by awarding me Grant no. NSERC A4823.

Above all, I wish to thank my wife Donna for all her encouragement and patience, without which this book could never have been written.

H. I. FREEDMAN

CONTENTS

Deterministic Mathematical Models in Population Ecology

PART I
INTRODUCTION

The first part of this book consists of two chapters, one on single-species growth and one on the topics related to predation and parasitism. There has been much theoretical and experimental work on both of these subjects. Many models have been suggested to explain experimental data, and many experiments have been performed and observations made in order to verify or disclaim particular models.

In addition to the interest in these subjects in their own right, in later parts of this book they are combined to yield predator-prey models and models describing competition.

Chapter 1
SINGLE-SPECIES GROWTH

1.1 INTRODUCTION

The dynamics of the growth of a population can be described if the functional behavior of the rate of growth is known. Of course, it is this functional behavior which is usually measured in the laboratory or in the field when the ecologist is interested in single-species population growth.

The usual assumption made is that the rate of growth is in some sense proportional to the number of species present. The proportionality "constant" may be dependent or independent of the number of the species present, and it may be dependent or independent of time. In general, when the model is time independent, the population growth may be described by the autonomous equation

$$x' = xg(x) \qquad (' = \frac{d}{dt}) \tag{1.1}$$

where x is the number of species present at time t. Throughout this chapter it is always assumed that $x \geq 0$. If the model is time dependent, the population growth may be given by the nonautonomous equation

$$x' = xh(x,t) \qquad (' = \frac{d}{dt}) \tag{1.2}$$

Models of the preceding type are suitable for the growth of populations where the numbers are large. In this case x may be taken to represent the population density or perhaps the biomass of the population. Models given by (1.1) or (1.2) are generally not suitable if the population numbers are small, such as in the situation described in Dixon and Cornwell (1970), which deals with an island population and the dynamics of about 600 moose and 22 wolves. The reason for this is that the differential equation models assume continuous birth and death rates, whereas in small populations that assumption is clearly false. Dixon and Cornwell found that the average birth rate among the wolves was one a year. In this case the natural model to describe the population dynamics is a difference equation, giving the change in population from one generation to the next. Hence an appropriate model for such a population is

$$x(t + 1) = x(t)g(x(t)) \tag{1.3}$$

where $x(t)$ is the population number in the t-th generation or appropriate time unit, and $x(t + 1)$ is the population number in the $(t + 1)$-st.

Finally, it should be noted that the population dynamics could be described by specifying the functional behavior of higher-order derivatives than the first. A second-order model given by

$$x'' = G(x,x') \tag{1.4}$$

will be considered.

All the preceding models are, as the title of this book implies, deterministic. Stochastic models, however, are also very important. For a review of stochastic models of single-species growth, the reader is referred to Pielou (1969).

1.2 MALTHUSIAN GROWTH

At this time the simplest case is considered, namely, the case where the species grows (i.e., birth minus death) at a constant rate times the number present, with no limitations on its resources.

Such a situation would be described by Eq. (1.1) with $g(x) \equiv k$, k a constant. The equation of growth is then

$$x' = kx \tag{1.5}$$

Let x_0 be the population at time $t = 0$. Then Eq. (1.5) integrates to

$$\ell n \frac{x}{x_0} = kt \tag{1.6}$$

or

$$x = x_0 e^{kt} \tag{1.7}$$

Such a population growth as described by Eq. (1.7) may be valid for a short time, but it clearly cannot go on forever. The first modification would then be to insert a correction term to account for the fact that resources are indeed limited. Such a model is described in the next section.

1.3 LOGISTIC GROWTH

In this section it is assumed that $g(x) = r[1 - x/K]$ which yields the model

$$x' = rx\left[1 - \frac{x}{K}\right] \tag{1.8}$$

This model simulates the following conditions: for small x the population behaves as in the Malthusian growth model, but for large x the members of the species compete with each other for the limited resources.

To solve differential Eq. (1.8), write as

$$\left(\frac{1}{x} + \frac{1}{K - x}\right) dx = r\ dt \tag{1.9}$$

Again, letting x_0 be the population at time zero, (1.9) integrates to

$$\ell n \left(\frac{x}{K - x} \frac{K - x_o}{x_o}\right) = rt$$

or

$$\frac{x}{K - x} = \frac{x_0}{K - x_0} e^{rt} \tag{1.10}$$

Solving (1.10) for x gives

$$x = \frac{Kx_0 e^{rt}}{K - x_0 + x_0 e^{rt}} \tag{1.11}$$

The population dynamics is then described as in Fig. 1.1. If $x_0 < K$, the population grows, approaching K asymptotically as $t \to \infty$. If $x_0 > K$, the population decreases, again approaching K asymptotically as $t \to \infty$. If $x_0 = K$, the population remains constant in time at $x = K$.

As is clearly seen in this model, K is a limiting factor on the growth of the population. It bounds above the size of populations which are initially small and tends to pull down populations which are initially large. K, of course, is itself a function of the resources (e.g., food, space, sunlight).

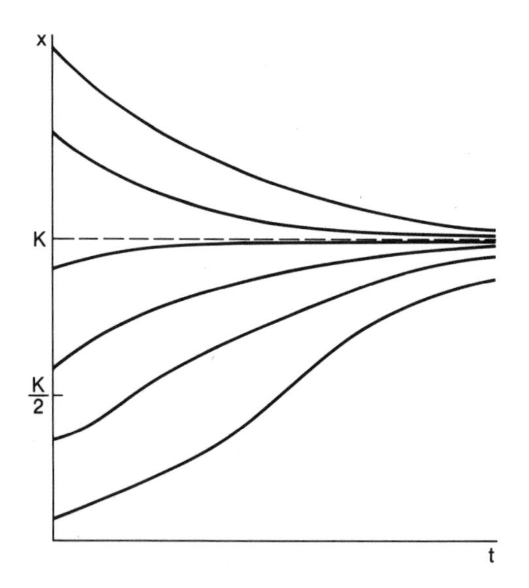

Fig. 1.1 Logistic growth. There is an inflection point at $x = K/2$ for solutions with $x_0 < K/2$.

Because many models which have been used by ecologists recently have the preceding properties, the logistic model is a prototype of a more general model to be discussed in the next section. The constant K is called the carrying capacity of the environment.

This model was utilized by Verhulst (1838) in his work on populations and similarly by Pearl (1930) and is sometimes referred to as the Verhulst-Pearl model.

Another aspect of single-species growth also began with this model. Here we mean the notions of r and K selection, where the r and K refer to their counterparts in (1.8). An r-selector is a species that grows in such a way as to increase its population quickly (i.e., so as to maximize r). A K-selector is a species which regulates its birth and death rates so as to maximize K.

1.4 THE GENERAL AUTONOMOUS MODEL

In this section Eq. (1.1) is analyzed under appropriate assumptions on $g(x)$, assumptions which have been used by ecologists in their work on population dynamics. These assumptions are as follows:

1. $g(x)$ is continuous with a piecewise continuous first derivative on $[0,\infty)$, and $g(0) > 0$.
2. $dg(x)/dx \leq 0$, $x \in [0,\infty)$, where $dg(x)/dx$ is understood to mean the right and left derivatives at points of discontinuity of the derivative.

Consider now Eq. (1.1) again. With the same meaning for x_0 as before, Eq. (1.1) integrates to

$$\int_{x_0}^{x} \frac{du}{ug(u)} = t \tag{1.12}$$

Under hypotheses 1 and 2 it is readily seen that $g(x)$ may be of one of two types:

1. $g(x) > 0$ for $x > 0$.
2. There exists a $K > 0$ such that $g(K) = 0$.

These two possibilities are examined separately. The first has as its prototype the Malthusian growth; the second has as its prototype the logistic model.

In case 1, since $g(u) > 0$, Eq. (1.12) clearly has a solution for positive t only if $x > x_0$. Also $\int_{x_0}^{x} du/ug(u)$ is a monotonically increasing function of x which tends to infinity as $x \to \infty$. Hence Eq. (1.12) always has a solution for $t > 0$, $x = x(t)$, such that $\lim_{t\to\infty} x(t) = \infty$, giving an unbounded population growth.

In case 2, without loss of generality, it may be assumed that

$$K = \sup \{x \mid g(x) > 0\}$$

Now since the left derivative of g(x) exists at $x = K$, in a left neighborhood of $x = K$, g(x) may be written as

$$g(x) = (x - K)^{\alpha} \tilde{g}(x) \tag{1.13}$$

where $\alpha \geq 1$ and $\lim_{x\to K^-} \tilde{g}(x) < \infty$. Consider now the case $x_0 < K$. Then, initially, $x > x_0$ in order that Eq. (1.12) have a solution for $t > 0$, and so long as $x_0 \leq x < K$, $\int_{x_0}^{x} du/ug(u)$ is monotonically increasing. Further, by (1.13)

$$\lim_{x\to K^-} \int_{x_0}^{x} \frac{du}{ug(u)} = \infty \tag{1.14}$$

Hence, from the monotonicity and (1.14), Eq. (1.12) always has a solution for $t \geq 0$, and this solution is bounded above by $x = K$.

If $x_0 > K$ and $g(x_0) \neq 0$ [and hence $g(x_0) < 0$], then setting

$$K_1 = \inf\{x \mid g(x) < 0\}$$

and arguing analogously to the case where $x_0 < K$, it is readily seen that once more Eq. (1.12) has a solution for $t \geq 0$ which decreases and is bounded below by $x = K_1$. If $x_0 \geq K$ and $g(x_0) = 0$, then one has directly from Eq. (1.1) that $x = x_0$ is the desired solution.

The dynamics of the population is given essentially by Fig. 1.1, Fig. 1.2, or Fig. 1.3, depending on where $g(x) = 0$.

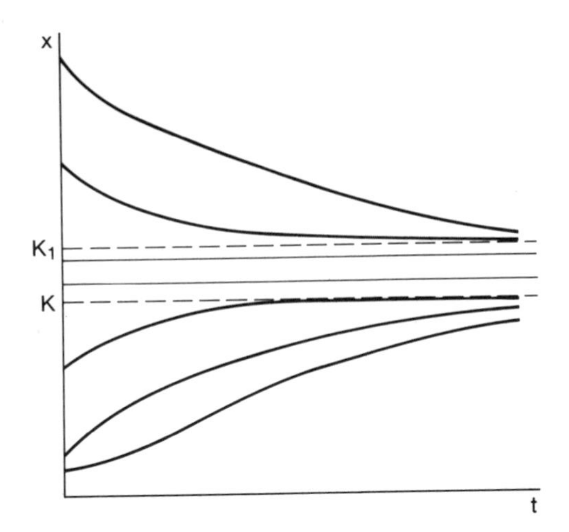

Fig. 1.2 Growth with carrying capacity, $K_1 > K$.

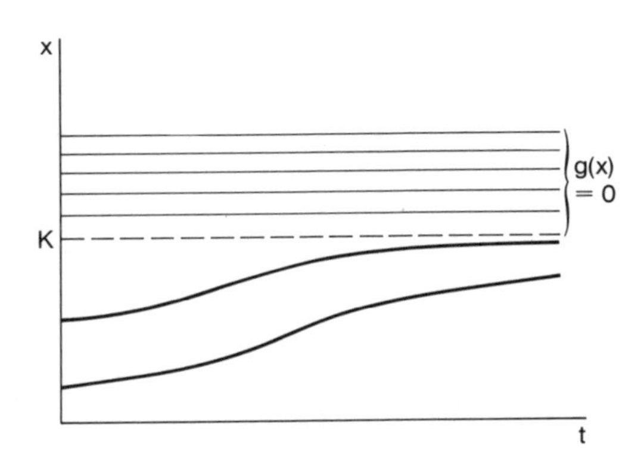

Fig. 1.3 Growth with carrying capacity, $g(x) = 0$ for $x \geq K$.

In most models as actually utilized by ecologists (see, for example, Schoener, 1973), $g(x)$ has a simple zero at $x = K$. Here again K is called the carrying capacity of the environment.

1.5 NONAUTONOMOUS GROWTH

In this section an attempt is made to separate the intrinsic growth factor of the species from its interaction growth factor. The interaction factor (and possibly the intrinsic factor) is assumed to be time dependent, and hence the model given by Eq. (1.2) is utilized. The ideas contained in this section are in the main due to Shock and Morales (1942).

Eq. (1.2) represents a time-dependent growth rate and will be written as

$$x' = xh(x,t), \qquad x(0) = x_0 \tag{1.15}$$

where $h(x,t)$ can be decomposed according to

$$h(x,t) = \alpha(x,t) + x^{k-1}\beta(x,t) \tag{1.16}$$

Here $\alpha(x,t)$ is the specific growth rate intrinsic to each member of the species regardless of the proximity of any other members of the species. The term $x^{k-1}\beta(x,t)$ is the specific growth rate due to interaction of one member of the species with its neighbors (usually due to competition but possibly due to reproduction as well), where $k \geq 1$ is a constant. Putting (1.15) into (1.16) gives as the model

$$x' = x\alpha(x,t) + x^k\beta(x,t), \qquad x(0) = x_0 \tag{1.17}$$

Under appropriate assumptions on $\alpha(x,t)$ and $\beta(x,t)$, it may be possible to analyze the model.

As an example, consider the following assumptions on $\alpha(x,t)$ and $\beta(x,t)$. First, for $\alpha(x,t)$ it is assumed that intrinsically each member of the species, if left alone, would have a constant specific growth rate, i.e., $\alpha(x,t) \equiv \alpha$, a constant. For $\beta(x,t)$ it is assumed that there is competition which is given by some function of t, $-f(t)$, that is, $\beta(x,t) \equiv -f(t)$. Further, it is assumed that this competition is proportional to the population, that is, $k = 2$. Putting these assumptions into Eq. (1.17) gives as a model

$$x' = \alpha x - f(t)x^2, \qquad x(0) = x_0 \tag{1.18}$$

Eq. (1.18) is a Bernoulli equation, which may be solved by letting $y = 1/x$. If $y_0 = 1/x_0$, Eq. (1.18) becomes

$$y' = -\alpha y + f(t), \qquad y(0) = y_0$$

which integrates to

$$y = e^{-\alpha t}\left[y_o + \int_o^t e^{\alpha s} f(s)\, ds\right] \tag{1.19}$$

or substituting back for x,

$$x = \frac{x_o e^{\alpha t}}{1 + x_o \int_o^t e^{\alpha s} f(s)\, ds} \tag{1.20}$$

Once $f(t)$ is given explicitly, the behavior of the population may be explicitly determined.

In case the population as modeled by Eq. (1.20) [or more generally by (1.17)] is bounded above asymptotically, there will again be a carrying capacity which this time is not a consequence of the environment only, but is also a consequence of the intra-competition of the species.

1.6 DISCRETE GROWTH

When the population remains small over a number of generations or remains essentially constant over a generation, it would seem that the dynamics of the population is best described by a discrete model.

A general model is given by Eq. (1.3), $x(t + 1) = x(t)g(x(t))$, where $x(t)$ is the population (or population density) of the t-th generation. Since one would expect the population of the next generation to be the population of the present generation plus the number born minus the number of deaths, the specific growth rate $g(x(t))$ may be taken to have the form

$$g(x(t)) = 1 + b(x(t)) - d(x(t)) \tag{1.21}$$

where $b(x(t))$ and $d(x(t))$ are the birth and death rates, respectively.

At this time, specific assumptions will be made concerning the birth and death rates. These assumptions, of course, are not relevant to all single-species ecosystems modeled by Eq. (1.3); they are based on the work done on specific species and would seem to be valid for a variety of such ecosystems. In particular, one may think of a host insect species whose life cycle occurs over a period of one year, whose birth rate is relatively constant, and whose death rate before oviposition (depositing eggs, after which the insect usually dies anyway) is controlled by a parasite; or one may think of a prey in an isolated environment, whose number is relatively small, whose birth rate is relatively constant, and whose death rate is controlled by a predator, which lives off that prey. For such systems, the following assumptions would seem to be reasonable:

1. $b(x(t)) = b$, a constant.
2. $d(x(t)) = \tilde{x}(t)/x(t)$, where $\tilde{x}(t)$ is the number attacked by the parasite or predator, as the case may be. The assumption is made that if a host insect is attacked by a parasite, it will die before it has a chance to reproduce; or if a prey is attacked by predators, it will be killed.

The model now takes the form

$$x(t + 1) = (1 + \hat{b})x(t) - \tilde{x}(t) \tag{1.22}$$

where $\hat{b}$, the adjusted birth rate, is the same as b for small numbers of isolated prey, but is equal to b - 1 for parasited insects to account for the fact that all the insects of the present generation are dead before the beginning of the new life cycle.

If a suitable formula could be obtained for $\tilde{x}(t)$, the model as now given by Eq. (1.22) could be analyzed. Such a formula has been determined by Watt in his work with insects. Watt's formula was also used in Dixon and Cornwell (1970) for predator-prey systems of the type described earlier, and so it seems appropriate to utilize it here. The reader is referred to Watt (1959) for a derivation of this formula [Eq. (1.23)] and the experimental evidence to justify it.

Based on the preceding, it is assumed that

$$\tilde{x}(t) = k\left(1 - e^{-cx(t)^2}\right) \tag{1.23}$$

where k and c are constants which depend on the number and properties of the parasites or predators, as the case may be. There have been modifications of Eq. (1.23). In particular, k and c can be made time dependent. For such a formula see Dixon and Cornwell (1970) or the Exercises. For the purposes of this analysis, however, Eq. (1.23) will suffice. Substituting for $\tilde{x}(t)$ in (1.22) now gives the model as

$$x(t + 1) = (1 + \hat{b})x(t) - k\left[1 - e^{-cx(t)^2}\right] \tag{1.24}$$

This model can now be analyzed.

For an insect, $\hat{b}$ would be very large, but then correspondingly k would be large as well. In fact, k would tend to be many times larger than $\hat{b}$. That would also be true for a prey, where $\hat{b}$ would be small, as, for example, in the situation described in Dixon and Cornwell (1970).

If $x(t) = k/\hat{b}$, the system is in a state of slow growth, for then $x(t + 1) = k/\hat{b} + ke^{-ck^2/\hat{b}^2}$, which is only slightly larger than $k/\hat{b}$ if $ke^{-ck^2/\hat{b}^2}$ is small (which is the case, for instance, in Dixon and Cornwell, 1970).

Suppose now that x(t) differs from $k/\hat{b}$ by a small amount, i.e.,

$$x(t) = \frac{k}{\hat{b}} + \varepsilon \tag{1.25}$$

where ε is so small in comparison with $k/\hat{b}$ that the approximation $e^{-c(k/\hat{b} + \varepsilon)^2} \approx e^{-ck^2/\hat{b}^2}$ is a reasonable one. Then from (1.24), $x(t + 1) = k/\hat{b} + (1 + \hat{b})\varepsilon + ke^{-ck^2/\hat{b}^2}$. Similarly, $x(t + 2) = (1 + \hat{b})\left[k/\hat{b} + (1 + \hat{b})\varepsilon + ke^{-ck^2/\hat{b}^2}\right] - k\left(1 - e^{-c\left[k/\hat{b} + (1 + \hat{b})\varepsilon + ke^{-ck^2/\hat{b}^2}\right]^2}\right) = k/\hat{b} + (1 + \hat{b})^2\varepsilon + [(1 + \hat{b})k + k]e^{-ck^2/\hat{b}^2}$, again assuming that $(1 + \hat{b})\varepsilon + ke^{-ck^2/\hat{b}^2}$ is small as compared to $k/\hat{b}$. Continuing in this manner, under the assumption that

$$(1 + \hat{b})^{n-1}\varepsilon + \frac{[(1 + \hat{b})^{n-1} - 1]}{\hat{b}} ke^{-ck^2/\hat{b}^2} << \frac{k}{\hat{b}} \tag{1.26}$$

then

$$x(t + n) = \frac{k}{\hat{b}} + (1 + \hat{b})^n + \frac{[(1 + \hat{b})^n - 1]}{\hat{b}} ke^{-k^2/\hat{b}^2} \tag{1.27}$$

i.e., so long as (1.26) remains valid, then at least for one more generation the growth will be relatively slow.

At this time the case where $x(t)$ is not close to $k/\hat{b}$ will not be considered, since then it is known that k (and c) will vary because the number of parasites or predators will vary. For such considerations a two-species predator-prey model is needed. Such a discussion occurs in Chapter 4.

Now we return to the general equation of discrete growth, i.e., Eq. (1.3). A general analysis of possible behaviors was given by Maynard Smith (1968) for the case in which $g(x)$ is as shown in Fig. 1.4, where x_1 and x_2 are those values of x where $g(x) = 1$. Clearly since $g(x_1) = 1$, $i = 1,2$, if at any time $x(t) = x_i$, then $x(t + 1) = x_i$, i.e., x_i, $i = 1, 2$, are steady states.

We will show here that x_1 is unstable and that x_2 is stable. First, if $x(t) < x_1$, then $g(x(t) < 1$ and $x(t + 1) < x(t) < x_1$. Then the sequence $x(t + n)$, $n = 1, 2, \ldots$, is decreasing and in the absence of lower steady states, the population goes extinct. If $x_1 < x(t) < x_2$, then $g(x(t)) > 1$ and $x(t + 1) > x(t)$. This means that populations initiating away from x_1, no matter how small the distance, continue to move away and x_1 is an unstable steady state.

From the preceding, if $x_1 < x(t) < x_2$, the population moves toward x_2. If $x(t) > x_2$, then $g(x(t)) < 1$ and $x(t + 1) < x(t)$ and the population again moves toward x_2. Hence x_2 is stable. The population may, of course, jump across x_2 and oscillations will then occur. It is even possible that the oscillations will be

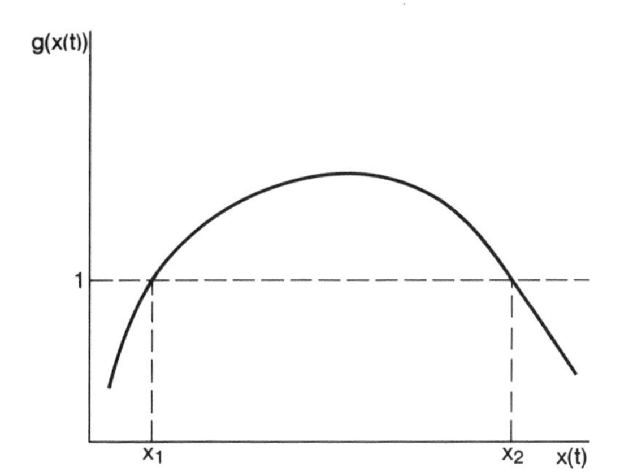

Fig. 1.4 A possible growth curve in the discrete case.

periodic of n generations. The question of periodic solutions and their bifurcations of discrete single-species systems has recently received much attention and has led to the definition of chaotic behavior of the solutions (see Li and Yorke, 1975, May, 1975, May and Oster, 1976).

1.7 SECOND-ORDER MODELS

It is conceivable that the growth pattern of a species may be better represented by a second-order (or even higher) differential equation than by a first-order equation. It has been pointed out in Clark (1971) that in some sense (comparing the dynamics of population growth to Newtonian mechanics), the second derivative may be thought of as representing the "life-force" of the species. For a suitable model Eq. (1.4) is used to begin with, i.e., $x'' = G(x,x')$. It is supposed that the "life-force" may be written as the sum of two quantities, a force of restoration to an equilibrium position (this is related to the existence of a carrying capacity of the environment), G_E, and a force due to the influence of the recent past history of the population, G_H. Consequently, $G(x,x')$ takes the form

$$G(x,x') = G_E + G_H \tag{1.28}$$

Since G_E is a restoration force, it is a function of x only. Hence G_E takes the form

$$G_E = -g(x) \tag{1.29}$$

where g(x) has the properties

1. There exists $x^* > 0$ (the equilibrium) such that $g(x^*) = 0$.
2. $(x - x^*)g(x) > 0$ for $x \neq x^*$ (the restoration effect).

G_H is to reflect the recent history of the population. It will be assumed that this recent history is influenced by the rate of growth of the population; i.e., G_H is proportional to x', with the "constant of proportionality" a function of the population. Hence G_H takes the form

$$G_H = -f(x)x' \tag{1.30}$$

Now combining (1.28) to (1.30), the model becomes

$$x'' + f(x)x' + g(x) = 0 \tag{1.31}$$

Equation (1.31) is a form of the Lienard equation, an equation which has been much discussed in the literature. In particular, it is well known (see the Exercises) that under suitable hypotheses on f(x), there will exist periodic solutions of Eq. (1.31), which oscillate about the equilibrium value $x = x^*$, unlike the first-order models, whose solutions approach their carrying capacities asymptotically. Here the population may have a "momentum," which takes it past the carrying capacity, whereupon the restoring force takes over to bring the population back. Specific forms of Eq. (1.31) are presented for analysis in the Exercises. The stability of the stationary solution and the periodic solutions when they exist may be determined in the usual manner (see Sec. D of the Appendix).

The main ideas in setting up this model are due to Clark (1971). Innis (1972) has pointed out that this model may be transformed into a first-order system, and as such is only a special case of such models. However, because so much is known about Eq. (1.31) because of its special form, it is worthwhile to be considered as it is. Equation (1.31) (with $x^* = 0$) is discussed in Birkhoff and Rota (1969), Cesari (1971), and La Salle and Lefschetz (1961).

NOTES ON THE LITERATURE

There has been much experimental work and field observation directed toward measuring single-species growth parameters. For papers along this line see Feller (1940), Gilpin (1974), Leslie (1957), Telfer (1971), and White and Huffaker (1969a). Other distinct types of models for single-species growth have also been proposed. Some of these are given in Hubbell (1973), Khanin and Dorfman (1973), Streifer and Istock (1973), Timin and Collier (1971), and Trubatch and Franco (1974).

Certain modifications of the basic models have been much discussed. Models incorporating time lags have been discussed by Bartlett (1957), Caperon (1969), May (1973d), May et al. (1974), Ross (1972), and Stirzaker (1975). Some papers involving age structure models are De Angelis (1975a), Dekker (1975), Gurtin and MacCamy (1974), Landahl and Hansen (1975), and Skellam et al. (1960). Some models have been proposed either as integral equations or as integro-differential equations. These include Brauer (1975), Cooke and Yorke (1972), Cushing (1976b), Mazanov (1973), Roughgarden (1974e), and Shilepsky (1974). Brauer and Sanchez (1975a,b) have analyzed a model incorporating harvesting.

Stability of discrete models has also been much discussed. Some of these results may be found in Beddington (1974), Hassell (1975), Leon (1975) (where the stability of periodic oscillations is analyzed), May (1974b, 1975), and van den Driessche (1974). There has, of course, been considerable work done on stochastic models. Some relevant papers are Barclay (1975), Bartlett et al. (1960), Deistler and Feichtinger (1974), Feldman and Roughgarden (1975), Keiding (1975), Kiester and Barakat (1974), Levins (1968), Lewontin and Cohen (1969), Montroll (1972), Powers and Lackey (1975), Salt (1966), Solomon (1968), and Tognetti (1975).

For some other aspects of single-species growth, MacArther (1968a) has indicated how selection could change stability or fitness, Levine (1975) has considered enrichment, Keiding (1975) and Miller and Botkin (1974) have considered extinction problems. Coulman et al. (1971) have given a systems approach to modeling.

EXERCISES

1.1 Analyze the autonomous growth model with sources given by $x' = xg(x) + r$, where r is a constant. How does r affect the growth in general and in particular what is its effect on the carrying capacity, if such would exist when $r = 0$?

1.2 What would be the growth behavior of a species modeled by Eq. (1.17) if $\alpha(x,t) \equiv \alpha$, a constant; $\beta(x,t) = -f(t)$; $k > 2$, a positive integer?

1.3 In Eq. (1.24) suppose that k is replaced by a function of t, k(t). Show that if k(t) is a slowly increasing function of t, it is still possible to obtain a slowly increasing population.

1.4 For Eq. (1.31)

(a) (Clark, 1971) Analyze the solutions if $f(x) \equiv a$ and $g(x) = b(x - x^*)$.

(b) Show that if $f(x) \equiv 0$, all solutions initially close to x^* are periodic.

(c) Show that there is a nontrivial periodic solution if $f(x) = a[(x - x^*)^2 - 1]$ and investigate its stability.

1.5 Let K(t) be a positive periodic function of t and consider the nonautonomous model $x' = xg(x,K(t))$, where g has the usual properties and $g(K(t),K(t)) = 0$. Show that this model with a periodically varying capacity always has a nontrivial periodic solution.

1.6 (Freedman and Waltman, 1975) The model $x_1' = \alpha_1 x_1(1 - x_1/K_1) - \varepsilon x_1 + \varepsilon x_2$, $x_2' = \alpha_2 x_2(1 - x_2/K_2) - \varepsilon x_2 + \varepsilon x_1$ can be viewed as a model of species growth in two habitats with dispersal. The term $1/\varepsilon$ can be thought of as the barrier strength. Show that as $\varepsilon \to +\infty$, the x_1 and x_2 equilibria approach a common value, and find this value.

Chapter 2
PREDATION AND PARASITISM

2.1 INTRODUCTION

This chapter is concerned with the functional dependence of one species on another, where the first species depends on the second for its food. Such a situation occurs when a predator lives off its prey or a parasite lives off its host. The functional dependence in general depends on many factors, for instance, the various species densities; the efficiency with which the predator can search out and kill the prey (or in the case of a parasite, its searching efficiency for a host); the handling time (i.e., the time the predator takes to eat the prey or the time the parasite needs to deposit eggs).

To make matters precise we formulate the following definitions:

DEFINITION 2.1 *Predation* is a mode of life in which food is primarily obtained by killing and consuming organisms. A *predator* is an organism that depends on predation for its food. A *prey* is an organism that is or may be seized by a predator to be eaten.

DEFINITION 2.2 *Parasitism* is a relationship in which an organism of one kind lives in, on, or in intimate association with an organism of a second kind, at the expense of which it obtains food and possibly other benefits. A *parasite* is an organism of the first kind referred to above. A *host* is an organism of the second kind.

DEFINITION 2.3 *Predator functional response* to prey density refers to the change in the density of prey attacked per unit of time per predator as the prey density changes. The *predation curve* is the graph of the predator functional response versus the prey density.

As already mentioned, the predator functional response, and hence the predation curve, is a function of the predator and prey densities, among other things. The same is true of the parasite functional response.

DEFINITION 2.4 *Parasite functional response* to host density refers to the change per unit of time in the density of hosts attacked per parasite as the host density changes. The *parasitism curve* is the graph of the parasite functional response versus the host density.

It is the main purpose of this chapter to describe the functions and curves defined in Defs. 2.3 and 2.4, including those both theoretically and experimentally derived. Such functions and curves play a prominent role in the intermediate predator-prey models (see Chap. 4) and their applications (Chap. 6) as well as in certain competition models (Chap. 8).

2.2 DISCUSSION OF PARAMETERS

In this section the more important of the parameters involved in the predation or parasitism functions are discussed. The notation used here will be used throughout the remainder of this chapter. These parameters are summarized in Table 2.1.

1. *Prey density.* The prey (or host) density is the number of prey per unit area or volume. It will be denoted by the symbol x. If the change in prey density from one time unit to another is to be emphasized, then $x(t)$ is used to represent the prey at the present time, whereas $x(s)$ represents the prey at a future time (usually $s = t + 1$).

Table 2.1 Summary of Parameters Used in Predation and Parasitism

Parameter	Notation
Prey density	x
Predator density	y
Area of discovery	A
Unit area of discovery	Q
Mutual interference constant	m
Total search time	T
Handling time	T_h
Attack coefficient	a
Predator functional response	p

2. *Predator density.* The predator (or parasite) density is the number of predators per unit area or volume and is denoted by the symbol y. The terms $y(t)$ and $y(s)$ have similar meanings to $x(t)$ and $x(s)$ as defined in item 1.

3. *Area of discovery.* The area of discovery A is the average area that a predator (or parasite) searches for its prey (or hosts) in one unit of time. A is usually taken to be constant (see Hassell and May, 1973), but there are cases in which A has been shown to be a function or prey or host density (e.g., in Matsumoto and Huffaker, 1973a, b, 1974).

4. *Unit area of discovery.* Where the area of discovery depends on the predator or parasite density (see Hassell, 1972, Hassell and Varley, 1969), the unit area of discovery Q is defined by

$$Q = Ay^m \tag{2.1}$$

where m is the mutual interference constant (defined in item 5). Q is the area of discovery for a single predator or parasite. Equation (2.1) was proposed by Hassell and Varley (1969).

5. *Mutual interference constant.* The symbol m is defined by Eq. (2.1) and represents the degree to which predators or parasites interfere with each other in their search for prey or hosts, respectively. Hence $0 \leq m < 1$.
6. *Total search time.* The total search time T in the case of a predator is the total time available for searching out (including capturing) and then devouring its food; for a parasite it includes the time available for searching out its hosts and then ovipositing its eggs.
7. *Handling time.* The handling time T_h is the average time spent in handling the prey or hosts (i.e., the time spent in capturing and eating prey by predators and the time spent ovipositing eggs by a parasite, $T_h \geq 0$). See Salt and Willard (1971) for a discussion of handling time as a function of size of prey.
8. *Attack coefficient.* The attack coefficient a is the rate at which a predator can search out its prey or a parasite its host.
9. *Predator functional response.* The predator or parasite functional response is denoted by p(x).

2.3 SOME SIMPLE THEORETICAL MODELS OF FUNCTIONAL RESPONSE

The simplest model of functional response is obtained by assuming that in the time available for searching the total change in host density is the attack coefficient times the total search time times the prey or host density, i.e.,

$$\text{Total change in } x = aTx \tag{2.2}$$

Hence the predator or parasite response function p(x) is

$$p(x) = ax \tag{2.3}$$

Such a response was used by Lotka (1925), Volterra (1927), Nicholson (1933), and Nicholson and Bailey (1935) in their work.

The following modification was proposed in Holling (1959). Since the predator's search time is reduced by the act of handling its prey, the time available for searching is $T - T_h Tp(x)$. Substituting in (2.2) and setting equal to $T_p(x)$ gives

$$p(x) = \frac{ax}{1 + aT_h x} \tag{2.4}$$

The preceding models do not yet incorporate the predator or parasite density (i.e., the fact that the efficiency of each predator or parasite may be reduced because of interference has not yet been taken into account). In order to do this assume that

$$aT = A \tag{2.5}$$

Equation (2.5) would be reasonable, for example, where a parasite parasitizes all hosts within its area of discovery. Then using (2.1) gives

$$a = \frac{Q}{T} y^{-m} \tag{2.6}$$

Substituting (2.6) into (2.3) or (2.4) gives a functional response that is predator (or parasite) dependent as well as prey (or host) dependent.

The predation curves for these functional responses are given in the next section. They are just special cases of a general functional response utilized in the "intermediate" predator-prey and competition models and their applications.

2.4 THEORETICAL AND EXPERIMENTAL PREDATION AND PARASITISM CURVES

In this section both the predation and parasitism curves are referred to as the predation curve. This section is devoted to reproducing various types of predation curves found experimentally by many investigators. It will turn out that all these types are just special cases of a general predation curve defined in Chap. 4.

The first of these types is that of a straight line through the origin (see Fig. 2.1). This would include the graph of Eq. (2.3), and may be found, among other places, in Lotka (1925), Matsumoto and Huffaker (1974), Nicholson (1933), Nicholson and Bailey (1935), and Volterra (1927).

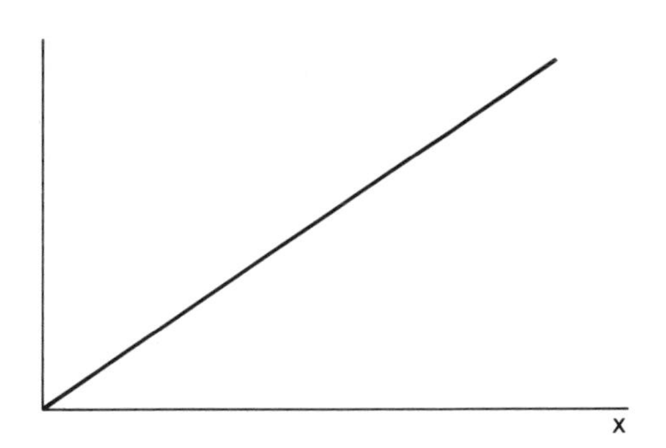

Fig. 2.1 Lotka-Volterra predation.

The second type of predation curve would be a curve emanating from the origin, having a negative concavity throughout and approaching a horizontal positive asymptote (see Fig. 2.2). This type of curve would include the graph of Eq. 2.4 and also can be found in Gause et al. (1936), Holling (1959), Matsumoto and Huffaker (1973b, 1974), Royama (1971); and elsewhere. (A curve is said to have a positive or negative concavity according to the sign of its second derivative.)

A similar type of predation curve is depicted in Fig. 2.3. The difference between this and the previous one is that here the curve is unbounded above. Such curves may be found in Holling (1959) and Royama (1971), for instance.

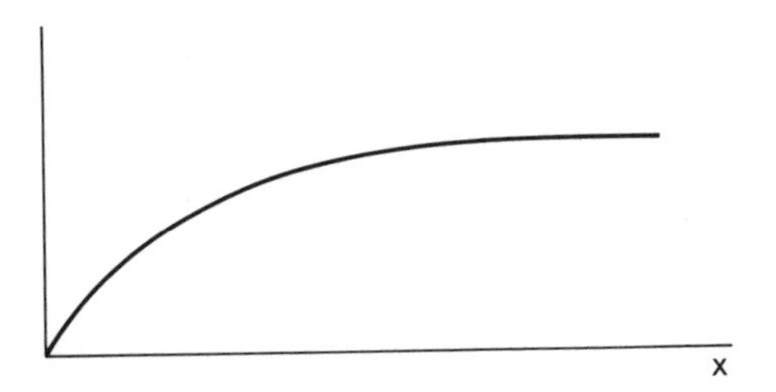

Fig. 2.2 Holling-type predation.

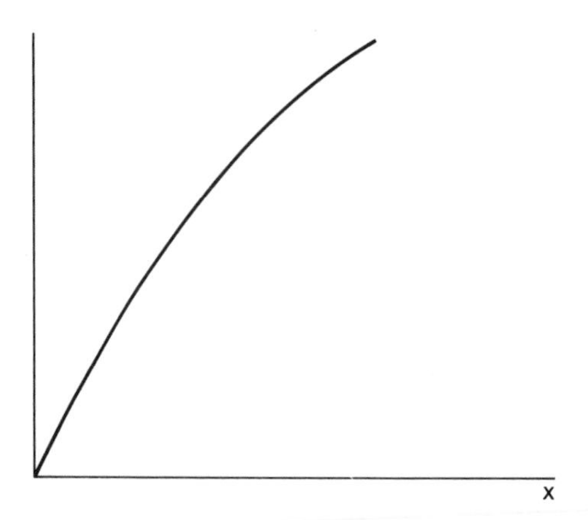

Fig. 2.3 Holling predation unbounded above.

The first of the sigmoid curves appears in Fig. 2.4. Here the predation curve experiences a change in concavity and is bounded above, again approaching a horizontal asymptote. Such curves may be found in Royama (1971).

The curve depicted in Fig. 2.5 again has a change in concavity, this time from negative to positive. It is also unbounded above. Such curves may be found in Holling (1959), Royama (1971), and Taylor (1974).

Finally, Fig. 2.6 depicts a curve with several concavity changes (i.e., a curve with several "s"-shapes). Such curves may be found in Haynes and Sisojevic (1966).

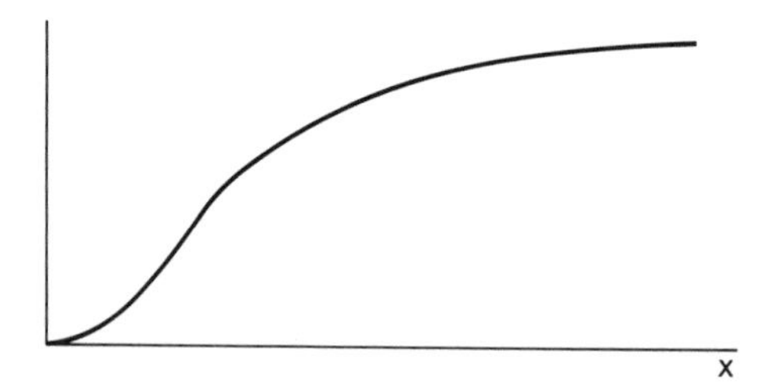

Fig. 2.4 Predation with a change in concavity, bounded above.

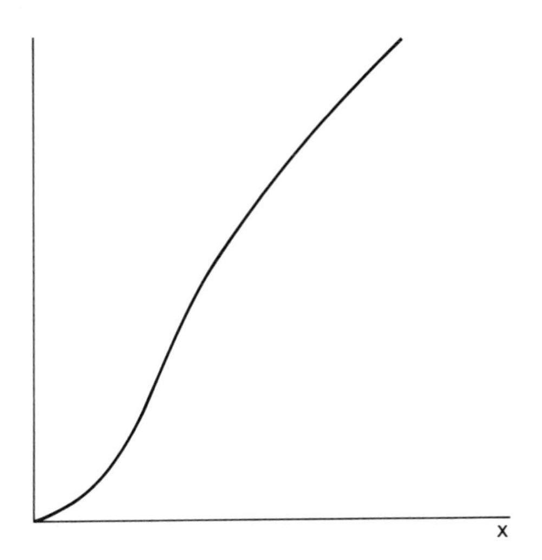

Fig. 2.5 Predation with a change in concavity, unbounded above.

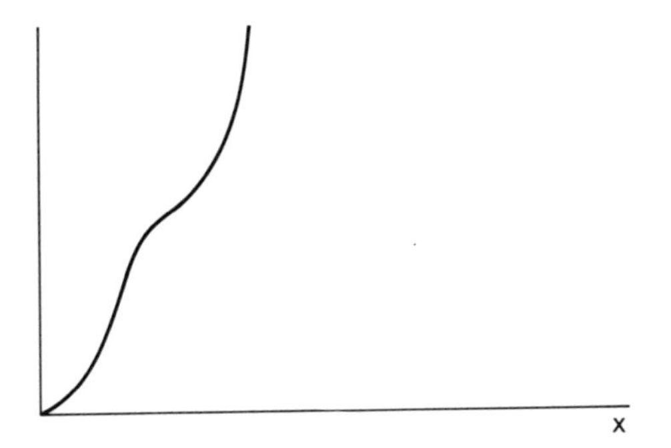

Fig. 2.6 Predation with several changes in concavity.

The common properties of all these predation curves that will be utilized later are that they

1. all pass through the origin
2. are all strictly increasing

2.5 OTHER ASPECTS OF PREDATION AND PARASITISM

Many authors and investigators have devoted much time and energy to other interesting aspects of predation and parasitism. However, since they are not germane (with a few exceptions) to what follows throughout this book, they will be mentioned only briefly. In particular, none of the formulas will be derived. However, references will be given whenever these formulas are discussed in detail.

One problem of concern to ecologists is to determine the number of survivors at the end of a time unit given the initial number of prey or hosts at the beginning of that interval. It would now help to think of x as the actual number of prey or hosts rather than the density, although the latter would be valid if one thinks of an initial density x(t) and a final density x(s) (the density of the survivors). If x is thought of as the number of prey or hosts, then y is to be thought of as the number of predators or parasites. A survey of formulas relating x(s) to x(t) is given in Hassell and May (1973), including several developed by those authors. The right-hand parts of these formulas are referred to as the "parasitism functions." (The paper deals mainly with parasitism as opposed to predation.) The following is based on that survey (Table 1 of Hassell and May, 1973).

1. $x(s) = x(t) \exp[-Ay(t)]$ (Nicholson, 1933, Nicholson and Bailey, 1935).

2. $x(s) = x(t) \exp\left[-\frac{aTy(t)}{1 + aT_h x(t)}\right]$ (Holling, 1959, Royama, 1971, Rogers, 1972).

3. $x(s) = x(t) \exp[-Qy(t)^{1-m}]$ (Hassell and Varley, 1969).

4. $x(s) = x(t) \exp\left[-\frac{Qy(t)^{1-m}}{1 + aT_h x(t)}\right]$ (Hassell and Rogers, 1972).

5. $x(s) = x(t) \sum_{i=1}^{n} \{\alpha_i \exp[-A\beta_i y(t)]\}$, α_i, β_i const. (Hassell and May, 1973).

6. $x(s) = x(t) \sum_{i=1}^{n} \left[\alpha_i \exp\{-Q[\beta_i y(t)]^{1-m}\}\right]$, α_i, β_i const. (Hassell and May, 1973).

Another problem of interest related to the one preceding is to determine the number of prey or hosts attacked. This is clearly $x(t) - x(s)$, and hence may be determined from the preceding formulas. Much work, however, has been done to derive formulas for $x(t) - x(s)$ directly. Some of these follow.

7. $x(t) - x(s) = y(t)K\left\{1 - \exp\left[-Qx(t)y(t)^{1-m}\right]\right\}$, K const. (Watt, 1959).

8. $x(t) - x(s) = y(t)f \exp[-dy(t)]\ 1\left\{- \exp\left[-Qx(t)y(t)^{1-m}\right]\right\}$, f, d const. (Watt, 1959).

9. $x(t) - x(s) = y(t)\left[\dfrac{ca\alpha y(t)^{-\beta}Tx(t)}{1 + caT_h x(t)}\right]$ where c is the eggs laid per parasite and α, β are constants (Hassell and Rogers, 1972).

NOTES ON THE LITERATURE

Some other papers describing predation or parasitism experiments leading to predation curves are Ables and Shepard (1974), Addicott (1974), Glen (1975), Hassell (1969), McMurtry and van de Vrie (1973), Salt (1967), Taylor (1974), White and Huffaker (1969a,b). Additional papers discussing mutual interference include Beddington (1975), Diamond (1967), and Rogers and Hassell (1974). Other aspects of predation and general comments may be found in Diamond (1974a), Holling (1968), Hubbell (1973), Lawton et al. (1975), Marten (1973), Murdoch and Marks (1973); Salt (1967); and Slobodkin (1974).

EXERCISE

2.1 Equations (2.3) and (2.4) are algebraic functions whose graphs are of the form given in Figs. 2.1 and 2.2, respectively. Find algebraic functions representative of the graphs in Figs. 2.3 to 2.6.

PART II
PREDATOR-PREY SYSTEMS

In the next four chapters, models simulating predator-prey interactions are discussed. In Chap. 3, the simplest predator-prey system, the Lotka-Volterra model, is analyzed. Modifications and perturbations are also looked at. In Chap. 4, the so-called intermediate models are discussed. Here the materials of Chaps. 1 and 2 are assimilated into the models. Discrete models are also analyzed. In Chap. 5, the most general growth model, the Kolmogorov-type model, is investigated. Finally, Chap. 6 deals with related topics (food chains and several prey) and applications (pest control and immunity).

Chapter 3

LOTKA-VOLTERRA SYSTEMS FOR PREDATOR-PREY INTERACTIONS

3.1 INTRODUCTION

In 1925, during a conversation with Vito Volterra, a young zoologist by the name of Umberto d'Ancona, Volterra's future son-in-law, pointed out that in the years following World War I the proportion of predator fish caught in the Upper Adriatic was up from before, whereas the proportion of prey fish was down. This was a phenomenon which seemed to be predicted by one of Volterra's models. From that time on, Volterra, who was by then a very prominent mathematician, devoted his studies to models in ecology. (For a nice treatment of Volterra's work in ecology, see Scudo, 1971.)

In the same year that Volterra became interested in ecology, A. J. Lotka published a book titled *Elements of Physical Biology*. In this text he discussed the same model utilized by Volterra for predator-prey interactions. It is safe to assume, of course, that the two were completely unaware of each other's work. It is this model that is now known as the Lotka-Volterra model.

As usual we will let $x(t)$ stand for the prey density (or number, or biomass) and $y(t)$ for the predator density (or number, or biomass). Then the same assumption will be made here as was made in Chap. 1, namely, that the growth rate of any species at a given time is proportional to the number of that species present at that time. Further general assumptions are that the species are living in a homogeneous environment and that age structures are not taken into account. With bisexual communities one can also consider only the female population, or an average reproduction rate considering both males and females.

For the prey species it is assumed that the prey growth, if left alone, is Malthusian, i.e., the specific growth rate is constant [as in Eq. (1.5)]. It is further assumed that the specific growth rate is diminished by an amount proportional to the predator density. This leads to the prey equation

$$x' = x(\alpha - \beta y), \qquad \alpha,\beta > 0 \tag{3.1a}$$

For the predator species it is assumed that in the absence of prey, the predators will become extinct exponentially (just as in radioactive decay) but that their growth rate is enhanced by an amount proportional to the prey density. This leads to the predator equation

$$y' = y(-\gamma + \delta x), \qquad \gamma,\delta > 0 \tag{3.1b}$$

System (3.1) is that system referred to as the Lotka-Volterra model for predator-prey interactions. It is in some sense the simplest model for such interactions. It certainly is the easiest model to analyze in that the solutions may be found explicitly in phase space. We do so in the next section. (Of course, we only consider cases where $x,y \geq 0$.)

3.2 SOLUTION OF THE LOTKA-VOLTERRA SYSTEM

We consider again system (3.1). Dividing Eq. (3.1b) by (3.1a) leads to the equation

$$\frac{dy}{dx} = \frac{y(-\gamma + \delta x)}{x(\alpha - \beta y)} \tag{3.2}$$

or

$$\frac{\alpha - \beta y}{y}\, dy = \frac{-\gamma + \delta x}{x}\, dx \tag{3.3}$$

Suppose we initiate our system at $x_0,y_0 > 0$. Then from (3.3)

$$\int_{y_0}^{y} \left(\frac{\alpha}{v} - \beta\right) dv = \int_{x_0}^{x} \left(\frac{-\gamma}{u} + \delta\right) du \tag{3.4}$$

or

$$\alpha \ \ell n \frac{y}{y_0} - \beta(y - y_0) = -\gamma \ \ell n \frac{x}{x_0} + \delta(x - x_0) \tag{3.5}$$

In order to examine Eq. (3.5) properly, we first write it as

$$\begin{aligned} \alpha \ \ell n \left(\frac{\beta y/\alpha}{\beta y_0/\alpha}\right) - \beta\left[\left(y - \frac{\alpha}{\beta}\right) - \left(y_0 - \frac{\alpha}{\beta}\right)\right] &= -\gamma \ \ell n \left(\frac{\delta x/\gamma}{\delta x_0/\gamma}\right) \\ &\quad + \delta\left[\left(x - \frac{\gamma}{\delta}\right) - \left(x_0 - \frac{\gamma}{\delta}\right)\right] \end{aligned} \tag{3.6}$$

or

$$\delta\left(x - \frac{\gamma}{\delta}\right) - \gamma \ \ell n \frac{\delta x}{\gamma} + \beta\left(y - \frac{\alpha}{\beta}\right) - \alpha \ \ell n \frac{\beta y}{\alpha} = c_o \tag{3.7}$$

where

$$c_o = \delta\left(x_0 - \frac{\gamma}{\delta}\right) - \gamma \ \ell n \frac{\delta x_0}{\gamma} + \beta\left(y_0 - \frac{\alpha}{\beta}\right) - \alpha \ \ell n \frac{\beta y_0}{\alpha}$$

We now effect the change of variables

$$w_1 = x - \frac{\gamma}{\delta}, \qquad w_2 = y - \frac{\alpha}{\beta} \tag{3.8}$$

which when substituted into (3.7) gives

$$\delta w_1 - \gamma \ \ell n\left(\frac{\delta w_1}{\gamma} + 1\right) + \beta w_2 - \alpha \ \ell n\left(\frac{\beta w_2}{\alpha} + 1\right) = c_o \tag{3.9}$$

At this point we let

$$\begin{aligned} \Phi_1(w_1) &= \delta w_1 - \gamma \ \ell n\left(\frac{\delta w_1}{\gamma} + 1\right) \\ \Phi_2(w_2) &= \beta w_2 - \alpha \ \ell n\left(\frac{\beta w_2}{\alpha} + 1\right) \end{aligned} \tag{3.10}$$

and note that

$$\frac{d\Phi_1(w_1)}{dw_1} = \frac{\delta^2 w_1}{\delta w_1 + \gamma}, \qquad \frac{d\Phi_2(w_2)}{dw_2} = \frac{\beta^2 w_2}{\beta w_2 + \alpha} \tag{3.11}$$

From (3.10) $\Phi_i(0) = 0$, $i = 1, 2$; from (3.11), $w_i[d\Phi_i(w_i)/dw_i] > 0$, $w_i \neq 0$, $i = 1, 2$, so long as $w_1 > -\gamma/\delta$ and $w_2 > -\alpha/\beta$, which by (3.8) is equivalent to $x, y > 0$. Further, $c_o > 0$, since

$c_o = \Phi_1[x_0-(\gamma/\delta)] + \Phi_2[y_0-(\alpha/\beta)]$. This means that Eq. (3.9), which can be written as

$$\Phi_1(w_1) + \Phi_2(w_2) = c_o \tag{3.12}$$

satisfies the criteria of Sec. D.6 of the Appendix for the existence of a periodic solution for arbitrary $x_0, y_0 > 0$, and so all solutions initiating in the first quadrant are periodic.

This includes as a special case when $x_0 = \gamma/\delta$, $y_0 = \alpha/\beta$, in which case $c_o = 0$, and hence corresponds to an equilibrium.

It only remains to examine the solutions initiating on the positive axes. If $x_0 > 0$, $y_0 = 0$, then $y \equiv 0$ and x satisfies

$$x' = \alpha x, \qquad x(0) = x_0 \tag{3.13}$$

which we have already solved, giving

$$x = x_0 e^{\alpha t} \tag{3.14}$$

Hence the x-axis is itself a solution with the flow away from the origin.

If $x_0 = 0$, $y_0 > 0$, then $x \equiv 0$ and y satisfies

$$y' = -\gamma y, \qquad y(0) = y_0 \tag{3.15}$$

Here the solution is

$$y = y_0 e^{-\gamma t} \tag{3.16}$$

which means the y-axis is also a solution with the flow toward the origin. The origin itself is, of course, an equilibrium, and by the preceding a hyperbolic point.

The preceding analysis is illustrated in Fig. 3.1.

As an interesting sidelight, it was thought that records kept by the Hudson Bay Company for the last 200 years seemed to confirm the general oscillatory behavior predicted by the Lotka-Volterra model. The records involved the catches of Canada lynx and its prey, the snowshoe hare. However, recently Gilpin (1973) has analyzed this data by computer and has found three reverse cycles.

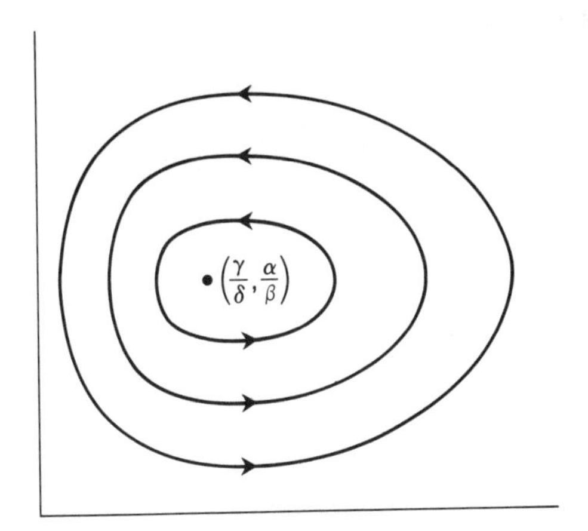

Fig. 3.1 Solutions in the phase plane for the Lotka-Volterra system (3.1).

If we did not know better, we might conclude that during these cycles the hare were eating the lynx. However, as pointed out by Gilpin, the most probable explanation is erratic trapper activity during these years.

3.3 INCREASING AND DIMINISHING RETURNS

System (3.1) was modified by Samuelson (1967) to incorporate what he called increasing and decreasing returns. These are achieved by considering the system

$$\begin{aligned} x' &= x(\alpha + c_1 x - \beta y) \\ y' &= y(-\gamma + \delta x + c_2 y) \end{aligned} \tag{3.17}$$

where, as before, $\alpha, \beta, \gamma, \delta > 0$. The case where c_1 and c_2 are both positive corresponds to the case of increasing returns, whereas the case where both are negative corresponds to diminishing returns. One could also talk about mixed returns when c_1 and c_2 have opposite signs, or semi-increasing or diminishing returns when either c_1 or c_2, but not both, is zero.

Biologically, increasing returns in either species means that to the second order, the growth of that species is enhanced by increasing the species density. Wolves may be able to hunt better as a pack, or buffalo defend themselves better as a group. Diminishing returns correspond to a situation where when a prey species gets too large it begins to hinder its own growth, or a predator species hinders its own growth to the second order. Basically, $c_1 < 0$ is equivalent to logistic growth in the prey species in the absence of predators.

We will first determine the equilibria of system (3.17). Of course (0,0) continues to be an equilibrium. The other equilibrium is determined by solving the system of equations

$$\begin{aligned} c_1x - \beta y &= -\alpha \\ \delta x + c_2y &= \gamma \end{aligned} \tag{3.18}$$

This system has a unique solution if and only if

$$c_1c_2 + \beta\delta \neq 0 \tag{3.19}$$

Clearly condition (3.19) could be violated only in the case of mixed returns. Then assuming (3.19) to hold, if x^* and y^* are the solutions of (3.18), Cramer's rule yields

$$x^* = \frac{\beta\gamma - \alpha c_2}{\beta\delta + c_1c_2}, \qquad y^* = \frac{\alpha\delta + \gamma c_1}{\beta\delta + c_1c_2} \tag{3.20}$$

In order to ensure positive equilibria, we will insist that when $c_2 > 0$, then $c_2 < \beta\gamma/\alpha$, and when $c_1 < 0$, then $-c_1 < \alpha\delta/\gamma$.

We could, of course, now determine the stability of this critical point (see Sec. D.3 of the Appendix). This would give, of course, only the local stability of the critical point. In system (3.17), however, we are able to obtain the global picture by examining an appropriate Liapunov function (see Sec. D.6 of the Appendix).

We first effect the change of variables

$$w_1 = x - x^*, \qquad w_2 = y - y^* \tag{3.21}$$

which will have the effect of shifting the equilibrium of interest to the origin. Substituting (3.21) into (3.17) and simplifying gives

$$\begin{aligned} w_1' &= (w_1 + x^*)(c_1 w_1 - \beta w_2) \\ w_2' &= (w_2 + y^*)(\delta w_1 + c_2 w_2) \end{aligned} \tag{3.22}$$

Now, taking our hint from Eq. (3.9), we let

$$V(w_1,w_2) = \delta w_1 - \delta x^* \ln \frac{w_1 + x^*}{x^*} + \beta w_2 - \beta y^* \ln \frac{w_2 + y^*}{y^*} \tag{3.23}$$

be our designated Liapunov function. We note that all solutions of $V(w_1,w_2) = c > 0$ for $w_1 > -x^*$ and $w_2 > -y^*$ are closed curves about the origin in the w_1-w_2 plane.

We now consider $\dot{V}(w_1,w_2)$, the derivative of V along solutions.

$$\begin{aligned} \dot{V}(w_1,w_2) &= \frac{\partial V}{\partial w_1} w_1' + \frac{\partial V}{\partial w_2} w_2' \\ &= \delta c_1 w_1^2 + \beta c_2 w_2^2 \end{aligned} \tag{3.24}$$

The analysis of increasing, decreasing, and semi-increasing or semidecreasing returns is now clear. Suppose we first have increasing or semi-increasing returns. Then $\dot{V} > 0$ for $w_1^2 + w_2^2 \neq 0$. This means that all solutions cross the curves

$$V(w_1,w_2) = c > 0 \tag{3.25}$$

from inside to outside. Hence the solutions are spirals flowing away from the origin in the w_1-w_2 plane as illustrated in Fig. 3.2a. Translating back to the x-y plane, the shape of the solutions remain the same as shown in Fig. 3.2b spiraling about (x^*,y^*).

For decreasing or semidecreasing returns, $\dot{V} < 0$ for $w_1^2 + w_2^2 \neq 0$. In this case all solutions cross the curves (3.25) from outside to inside, and the origin is a global attractor. The solutions are then spirals, spiraling into the origin, as shown in

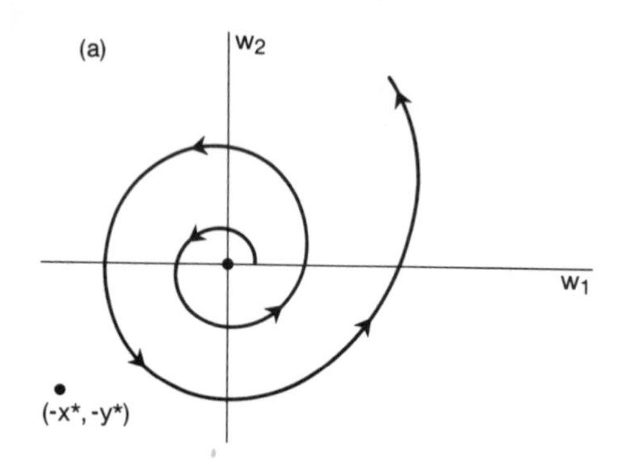

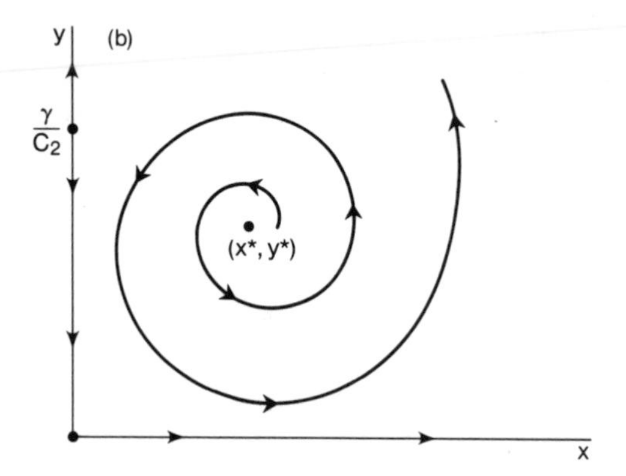

Fig. 3.2 (a) The case of increasing returns in the transposed plane. (b) The case of increasing returns in the original plane.

Fig. 3.3a. Translated into the x-y plane, then all solutions initiating in the first quadrant spiral into the equilibrium (x^*,y^*). This is shown in Fig. 3.3b.

We also consider the flows along the axes. Along the x-axis, $y \equiv 0$, and the flow along the x-axis is given by

$$x' = x(\alpha + c_1x) \tag{3.26}$$

In the case of increasing returns, the prey species increases without limit and the flow is away from the origin. In the case of decreasing returns, there is an equilibrium at $x = -\alpha/c_1$, and the flow is toward that equilibrium (the logistic flow of single-species growth).

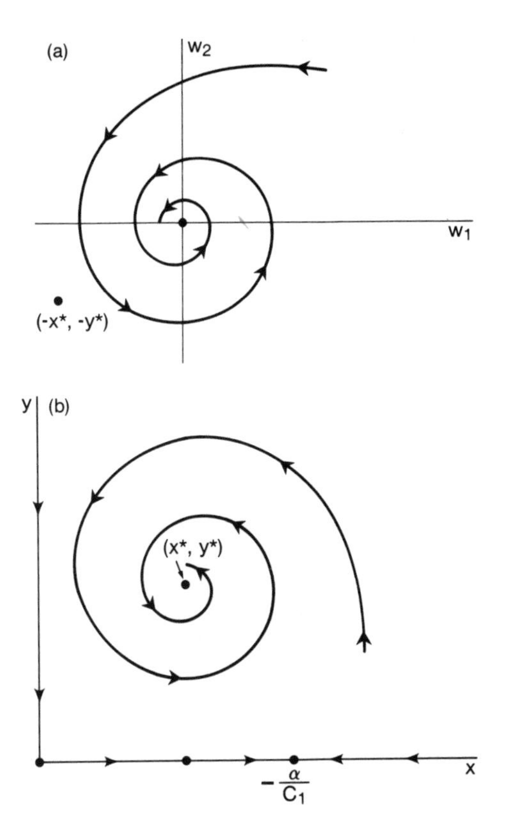

Fig. 3.3 (a) The case of decreasing returns in the transposed plane. (b) The case of decreasing returns in the original plane.

Along the y-axis, $x \equiv 0$ and the solutions are given by

$$y' = y(-\gamma + c_2 y) \tag{3.27}$$

If $c_2 > 0$ (increasing returns), there is an equilibrium at $y = \gamma/c_2$. In this case the equilibrium is unstable and the flow is away from it (see Fig. 3.2b). If $c_2 < 0$, the flow is entirely down the axis toward the origin.

The general case of mixed returns cannot be analyzed using (3.23), and the behavior of the solutions is considerably more complicated, but not of sufficient interest to be considered at this time.

3.4 PERTURBED MODELS

The Lotka-Volterra-type models discussed so far have all had the property of having either all solutions periodic or no solution other than the equilibria being periodic. The question has been posed by Samuelson (1971) as to whether the Lotka-Volterra system can be so modified as to yield a unique nonconstant periodic orbit.

Samuelson (1967) suggested that equations of the following type be considered:

$$\begin{aligned} x' &= x[\alpha + \varepsilon R_1(x) - \beta y] \\ y' &= y[-\gamma + \delta x + \varepsilon R_2(y)] \end{aligned} \tag{3.28}$$

where ε is a small positive parameter and $R_1(x)$, $R_2(y)$ are such that they cause increasing returns for values close to the equilibrium but diminishing returns far from equilibrium. A proper analysis of system (3.28) was not given by Samuelson, and it remains an open question as to what conditions on $R_1(x)$ and $R_2(y)$ do yield a *unique* periodic orbit.

More generally, Freedman and Waltman (1975a,b) have considered general perturbed Lotka-Volterra systems of the form

$$\begin{aligned} x' &= x(\alpha - \beta y) - \varepsilon f_1(x,y) \\ y' &= y(-\gamma + \delta x) - \varepsilon f_2(x,y) \end{aligned} \tag{3.29}$$

where ε is a small positive parameter. This is the most general perturbed autonomous ordinary differential equations model of the Lotka-Volterra system. The object is to find conditions on f_1 and f_2 which give stable limit cycles. This will be carried out by using Samuelson's suggestions precisely. First f_1 and f_2 will be determined so that close to the equilibrium we will have increasing returns, but away from the equilibrium we will have diminishing returns. Then using the Poincaré-Bendixson theorem (see Sec. E of the Appendix) we will be able to conclude the existence of a stable limit cycle.

We will see that this technique does not always work and that other perturbation techniques due to Loud (1959, 1964) are needed in certain deeper cases.

The analysis will be carried out in the next three sections.

3.5 THE PERTURBED EQUILIBRIUM

We consider now system (3.29) and note that for $\varepsilon = 0$, the equilibrium is given again by $(\gamma/\delta, \alpha/\beta)$. The question of the perturbed equilibrium reduces to checking the equations

$$\begin{aligned} \alpha x - \beta xy - \varepsilon f_1(x,y) &= 0 \\ -\gamma y + \delta xy - \varepsilon f_2(x,y) &= 0 \end{aligned} \tag{3.30}$$

for solutions of x and y as functions of ε close to γ/δ and α/β, respectively. We do this by utilizing the implicit function theorem (see Sec. B of the Appendix). We compute the Jacobian matrix for (3.30), denoted $J(x,y,\varepsilon)$, and obtain

$$J(x,y,\varepsilon) = \begin{bmatrix} \alpha - \beta y - \varepsilon f_{1x}(x,y) & -\beta x - \varepsilon f_{1y}(x,y) \\ \delta y - \varepsilon f_{2x}(x,y) & -\gamma + \delta x - \varepsilon f_{2y}(x,y) \end{bmatrix} \tag{3.31}$$

Hence

$$J\left(\frac{\gamma}{\delta}, \frac{\alpha}{\beta}, 0\right) = \begin{bmatrix} 0 & -\frac{\beta\gamma}{\delta} \\ \frac{\alpha\delta}{\beta} & 0 \end{bmatrix} \tag{3.32}$$

This gives $\det|J(\gamma/\delta,\alpha/\beta,0)| = \alpha\gamma \neq 0$, and so by the implicit function theorem, (3.30) can be solved for x and y as functions of ε for sufficiently small ε. Let these solutions be $x^*(\varepsilon)$ and $y^*(\varepsilon)$, respectively. We wish to compute x^* and y^* to order ε^2. Hence we set

$$\begin{aligned} x^* &= \frac{\gamma}{\delta} + x_1^*\varepsilon + x_2^*\varepsilon^2 + o(\varepsilon^2) \\ y^* &= \frac{\alpha}{\beta} + y_1^*\varepsilon + y_2^*\varepsilon^2 + o(\varepsilon^2) \end{aligned} \tag{3.33}$$

Substituting $x = x^*$ and $y = y^*$ into (3.30), expanding $f_i(x,y)$, $i = 1, 2$, in Taylor series about $\varepsilon = 0$, and utilizing (3.33) gives, after comparing the coefficients of ε and ε^2 and utilizing (3.34),

$$x_1^* = \frac{\beta}{\alpha\delta} f_2\left(\frac{\gamma}{\delta}, \frac{\alpha}{\beta}\right), \qquad y_1^* = -\frac{\delta}{\beta\gamma} f_1\left(\frac{\gamma}{\delta}, \frac{\alpha}{\beta}\right) \tag{3.34}$$

$$\begin{aligned} x_2^* &= \frac{\beta}{\alpha^2\gamma} f_1\left(\frac{\gamma}{\delta}, \frac{\alpha}{\beta}\right) f_2\left(\frac{\gamma}{\delta}, \frac{\alpha}{\beta}\right) - \frac{1}{\alpha\gamma} f_1\left(\frac{\gamma}{\delta}, \frac{\alpha}{\beta}\right) f_{2y}\left(\frac{\gamma}{\delta}, \frac{\alpha}{\beta}\right) \\ &\quad + \frac{\beta^2}{\alpha^2\delta^2} f_2\left(\frac{\gamma}{\delta}, \frac{\alpha}{\beta}\right) f_{2x}\left(\frac{\gamma}{\delta}, \frac{\alpha}{\beta}\right) \\ y_2^* &= \frac{\delta}{\alpha\gamma^2} f_1\left(\frac{\gamma}{\delta}, \frac{\alpha}{\beta}\right) f_2\left(\frac{\gamma}{\delta}, \frac{\alpha}{\beta}\right) + \frac{\delta^2}{\beta^2\gamma^2} f_1\left(\frac{\gamma}{\delta}, \frac{\alpha}{\beta}\right) f_{1y}\left(\frac{\gamma}{\delta}, \frac{\alpha}{\beta}\right) \\ &\quad - \frac{1}{\alpha\gamma} f_2\left(\frac{\gamma}{\delta}, \frac{\alpha}{\beta}\right) f_{1x}\left(\frac{\gamma}{\delta}, \frac{\alpha}{\beta}\right) \end{aligned} \tag{3.35}$$

Utilizing these last results, we shall examine the stability of (x^*,y^*) (see Sec. D of the Appendix). First we compute the variational matrix at (x^*,y^*), and we note that this matrix is just $J(x^*,y^*,\varepsilon)$ as given by (3.31). For convenience we write this matrix as

$$J(x^*,y^*,\varepsilon) = \begin{bmatrix} a_{11} & a_{12} \\ a_{21} & a_{22} \end{bmatrix} \tag{3.36}$$

Then the eigenvalues of $J(x^*,y^*,\varepsilon)$, $\lambda(\varepsilon)$, are given by

$$\lambda(\varepsilon) = \frac{1}{2}(a_{11} + a_{22}) \pm \frac{1}{2}\left[(a_{11} - a_{22})^2 + 4a_{12}a_{21}\right]^{1/2} \tag{3.37}$$

Utilizing (3.32), (3.33), (3.34), and (3.35), we first expand the a_{ij} in terms of ε to obtain

$$\begin{aligned} a_{11} &= \left[\frac{\delta}{\gamma} f_1\left(\frac{\gamma}{\delta}, \frac{\alpha}{\beta}\right) - f_{1x}\left(\frac{\gamma}{\delta}, \frac{\alpha}{\beta}\right)\right]\varepsilon + \left[-\frac{\beta\delta}{\alpha\gamma^2} f_1\left(\frac{\gamma}{\delta}, \frac{\alpha}{\beta}\right) f_2\left(\frac{\gamma}{\delta}, \frac{\alpha}{\beta}\right)\right. \\ &\quad - \frac{\delta^2}{\beta\gamma^2} f_1\left(\frac{\gamma}{\delta}, \frac{\alpha}{\beta}\right) f_{1y}\left(\frac{\gamma}{\delta}, \frac{\alpha}{\beta}\right) + \frac{\beta}{\alpha\gamma} f_2\left(\frac{\gamma}{\delta}, \frac{\alpha}{\beta}\right) f_{1x}\left(\frac{\gamma}{\delta}, \frac{\alpha}{\beta}\right) \\ &\quad \left. + \frac{\delta}{\beta\gamma} f_1\left(\frac{\gamma}{\delta}, \frac{\alpha}{\beta}\right) f_{1xy}\left(\frac{\gamma}{\delta}, \frac{\alpha}{\beta}\right) - \frac{\beta}{\alpha\delta} f_2\left(\frac{\gamma}{\delta}, \frac{\alpha}{\beta}\right) f_{1xx}\left(\frac{\gamma}{\delta}, \frac{\alpha}{\beta}\right)\right]\varepsilon^2 \\ &\quad + o(\varepsilon^2) \end{aligned} \tag{3.38}$$

$$a_{12} = -\frac{\beta\gamma}{\delta} + \left[-\frac{\beta^2}{\alpha\delta} f_2\left(\frac{\gamma}{\delta}, \frac{\alpha}{\beta}\right) - f_{1y}\left(\frac{\gamma}{\delta}, \frac{\alpha}{\beta}\right)\right]\varepsilon + o(\varepsilon)$$

$$a_{21} = \frac{\alpha\delta}{\beta} + \left[-\frac{\delta^2}{\beta\gamma} f_1\left(\frac{\gamma}{\delta}, \frac{\alpha}{\beta}\right) - f_{2x}\left(\frac{\gamma}{\delta}, \frac{\alpha}{\beta}\right)\right]\varepsilon + o(\varepsilon)$$

$$\begin{aligned} a_{22} = {} & \left[\frac{\beta}{\alpha} f_2\left(\frac{\gamma}{\delta}, \frac{\alpha}{\beta}\right) - f_{2y}\left(\frac{\gamma}{\delta}, \frac{\alpha}{\beta}\right)\right]\varepsilon + \left[\frac{\beta\delta}{\alpha^2\gamma} f_1\left(\frac{\gamma}{\delta}, \frac{\alpha}{\beta}\right) f_2\left(\frac{\gamma}{\delta}, \frac{\alpha}{\beta}\right)\right. \\ & - \frac{\delta}{\alpha\gamma} f_1\left(\frac{\gamma}{\delta}, \frac{\alpha}{\beta}\right) f_{2y}\left(\frac{\gamma}{\delta}, \frac{\alpha}{\beta}\right) + \frac{\beta^2}{\alpha^2\delta} f_2\left(\frac{\gamma}{\delta}, \frac{\alpha}{\beta}\right) f_{2x}\left(\frac{\gamma}{\delta}, \frac{\alpha}{\beta}\right) \\ & \left. + \frac{\delta}{\beta\gamma} f_1\left(\frac{\gamma}{\delta}, \frac{\alpha}{\beta}\right) f_{2yy}\left(\frac{\gamma}{\delta}, \frac{\alpha}{\beta}\right) - \frac{\beta}{\alpha\delta} f_2\left(\frac{\gamma}{\delta}, \frac{\alpha}{\beta}\right) f_{2xy}\left(\frac{\gamma}{\delta}, \frac{\alpha}{\beta}\right)\right]\varepsilon^2 \\ & + o(\varepsilon^2) \end{aligned}$$

From this it is clear that

$$(a_{11} - a_{22})^2 + 4a_{12}a_{21} = -4\alpha\gamma + o(1) \tag{3.39}$$

and so for sufficiently small ε, (x^*, y^*) is either a center or a spiral point and

$$\operatorname{Re} \lambda(\varepsilon) = \frac{1}{2}(a_{11} + a_{22}) \tag{3.40}$$

or, using (3.38),

$$\begin{aligned} \operatorname{Re} \lambda(\varepsilon) = {} & \left[\frac{\delta}{\gamma} f_1\left(\frac{\gamma}{\delta}, \frac{\alpha}{\beta}\right) + \frac{\beta}{\alpha} f_2\left(\frac{\gamma}{\delta}, \frac{\alpha}{\beta}\right) - f_{1x}\left(\frac{\gamma}{\delta}, \frac{\alpha}{\beta}\right) - f_{2y}\left(\frac{\gamma}{\delta}, \frac{\alpha}{\beta}\right)\right]\varepsilon \\ & + \left\{\frac{\beta\delta}{\alpha\gamma}\left(\frac{1}{\alpha} - \frac{1}{\gamma}\right) f_1\left(\frac{\gamma}{\delta}, \frac{\alpha}{\beta}\right) f_2\left(\frac{\gamma}{\delta}, \frac{\alpha}{\beta}\right)\right. \\ & + \frac{\delta}{\gamma} f_1\left(\frac{\gamma}{\delta}, \frac{\alpha}{\beta}\right)\left[-\frac{\delta}{\beta\gamma} f_{1y}\left(\frac{\gamma}{\delta}, \frac{\alpha}{\beta}\right) + \frac{1}{\alpha} f_{2y}\left(\frac{\gamma}{\delta}, \frac{\alpha}{\beta}\right)\right. \\ & \left. - \frac{1}{\beta} f_{1xy}\left(\frac{\gamma}{\delta}, \frac{\alpha}{\beta}\right) - \frac{1}{\beta} f_{2yy}\left(\frac{\gamma}{\delta}, \frac{\alpha}{\beta}\right)\right] + \frac{\beta}{\alpha} f_2\left(\frac{\gamma}{\delta}, \frac{\alpha}{\beta}\right) \\ & \left[\frac{1}{\gamma} f_{1x}\left(\frac{\gamma}{\delta}, \frac{\alpha}{\beta}\right) + \frac{\beta}{\alpha\delta} f_{2x}\left(\frac{\gamma}{\delta}, \frac{\alpha}{\beta}\right) - \frac{1}{\delta} f_{1xx}\left(\frac{\gamma}{\delta}, \frac{\alpha}{\beta}\right)\right. \\ & \left.\left. - \frac{1}{\delta} f_{2xy}\left(\frac{\gamma}{\delta}, \frac{\alpha}{\beta}\right)\right]\right\}\varepsilon^2 + o(\varepsilon^2) \\ = {} & [\quad]_1\varepsilon + [\quad]_2\varepsilon^2 + o(\varepsilon^2) \end{aligned} \tag{3.41}$$

We now suppose for the sake of definiteness that $\varepsilon > 0$. Then taking (3.41) as the definition of $[\ ]_1$, $[\ ]_2$, etc., we have proved the following:

1. If $[\ ]_1 > 0$, then the equilibrium (x^*,y^*) for sufficiently small $\varepsilon > 0$ is an unstable spiral point.
2. If $[\ ]_1 < 0$, then (x^*,y^*) is a stable spiral point.
3. If $[\ ]_1 = 0$, $[\ ]_2 > 0$, then (x^*,y^*) is an unstable spiral point.
4. If $[\ ]_1 = 0$, $[\ ]_2 < 0$, then (x^*,y^*) is a stable spiral point.

We remark that if $[\ ]_1 = [\ ]_2 = 0$, then by considering expansions to higher powers of ε, we can determine $[\ ]_3$, $[\ ]_4$, etc., and if one of these higher brackets is not zero, then the stability of the equilibrium is stable or unstable as the first nonvanishing bracket is negative or positive. If all brackets vanish, then the linear variation does not determine the stability.

3.6 CONSTANT AND LINEAR PERTURBATION

We first consider the case that $f_1\left(\frac{\gamma}{\delta}, \frac{\alpha}{\beta}\right)^2 + f_2\left(\frac{\gamma}{\delta}, \frac{\alpha}{\beta}\right)^2 \neq 0$. Then the equilibrium is indeed perturbed and the analysis of the previous section holds. We refer to this as the constant perturbation case. We shall be interested in those conditions on f_1 and f_2 as presented in the previous section which lead to an unstable spiral point, or increasing returns near the equilibrium.

To study the system properly, we first effect the change of variables

$$w_1 = x - x^*, \qquad w_2 = y - y^* \tag{3.42}$$

After some simplification on substituting into system (3.29), we get the system

$$\begin{aligned} w_1' &= (\alpha - \beta y^*)w_1 - \beta x^* w_2 - \beta w_1 w_2 \\ &\quad - \varepsilon[f_1(w_1 + x^*, w_2 + y^*) - f_1(x^*,y^*)] \end{aligned} \tag{3.43}$$

$$w_2' = \delta y^* w_1 + (\delta x^* - \gamma)w_2 + \delta w_1 w_2$$
$$- \varepsilon[f_2(w_1 + x^*, w_2 + y^*) - f_2(x^*,y^*)]$$

So as to sort out all the linear terms, we introduce the functions

$$\begin{aligned} g_1(w_1,w_2) &= f_1(w_1 + x^*, w_2 + y^*) - f_1(x^*,y^*) - \kappa w_1 - \lambda w_2 \\ g_2(w_1,w_2) &= f_2(w_1 + x^*, w_2 + y^*) - f_2(x^*,y^*) - \mu w_1 - \nu w_2 \end{aligned} \tag{3.44}$$

where

$$\begin{aligned} \kappa &= f_{1x}(x^*,y^*), \qquad \lambda = f_{1y}(x^*,y^*) \\ \mu &= f_{2x}(x^*,y^*), \qquad \nu = f_{2y}(x^*,y^*) \end{aligned} \tag{3.45}$$

Further, we set

$$\begin{aligned} b_{11} &= \alpha - \beta y^* - \varepsilon\kappa, \qquad b_{12} = -\beta x^* - \varepsilon\lambda \\ b_{21} &= \delta y^* - \varepsilon\mu, \qquad b_{22} = \delta x^* - \gamma - \varepsilon\nu \end{aligned} \tag{3.46}$$

Then system (3.45) may be written

$$\begin{aligned} w_1' &= b_{11}w_1 + b_{12}w_2 - \beta w_1 w_2 - \varepsilon g_1(w_1,w_2) \\ w_2' &= b_{21}w_1 + b_{22}w_2 + \delta w_1 w_2 - \varepsilon g_2(w_1,w_2) \end{aligned} \tag{3.47}$$

To analyze system (3.47), we again choose an appropriate Liapunov function, similar to the one chosen in Sec. 3.3. We let

$$V(w_1,w_2) = \Phi_1(w_1) + \Phi_2(w_2) \tag{3.48}$$

where

$$\Phi_1(w_1) = \frac{1}{\beta}\left[w_1 + \frac{b_{12}}{\beta}\ln\left|\frac{b_{12} - \beta w_1}{b_{12}}\right|\right] \tag{3.49}$$

and

$$\Phi_2(w_2) = \frac{1}{\delta}\left[w_2 - \frac{b_{21}}{\delta}\ln\left|\frac{b_{21} + w_2}{b_{21}}\right|\right] \tag{3.50}$$

We further let

$$\Gamma_c = \{(w_1,w_2)\,|\,V(w_1,w_2) = c\} \tag{3.51}$$

for fixed $c > 0$. Under further restrictions on the functions f_1 and f_2, we will show that given sufficiently small $\varepsilon > 0$ and sufficiently large c, all solutions of (3.47) cross Γ_c from outside to inside, and hence the system for large values of $w_1^2 + w_2^2$ would be in a state of decreasing returns.

Given a Γ_c, it was shown in Freedman and Waltman (1975a) that for sufficiently small ε,

$$\beta w_1 - b_{12} \geq \tfrac{1}{4}\beta d_1(0) > 0, \qquad \delta w_2 + b_{21} \geq \tfrac{1}{4}\delta d_2(0) \tag{3.52}$$

on Γ_c.

Now we compute $\dot{V}$, the derivative of V along solutions of system (3.47),

$$\begin{aligned}\dot{V}(w_1,w_2) &= \frac{b_{11}w_1^2}{\beta w_1 - b_{12}} + \frac{b_{22}w_2^2}{\delta w_2 + b_{21}} \\ &\quad - \varepsilon\left[\frac{w_1}{\beta w_1 - b_{12}}\, g_1(w_1,w_2) + \frac{w_2}{\delta w_2 + b_{21}}\, g_2(w_1,w_2)\right]\end{aligned} \tag{3.53}$$

Using (3.46), (3.45), (3.34), and (3.44), one obtains that

$$\begin{aligned}b_{11}w_1^2 - \varepsilon w_1 g_1(w_1,w_2) &= \left[\frac{\delta}{\gamma}\, f_1\!\left(\frac{\gamma}{\delta}, \frac{\alpha}{\beta}\right)w_1^2 + f_1\!\left(\frac{\gamma}{\delta}, \frac{\alpha}{\beta}\right)w_1\right. \\ &\quad + f_{1y}\!\left(\frac{\gamma}{\delta}, \frac{\alpha}{\beta}\right)w_1w_2 \\ &\quad \left. - w_1f_1\!\left(w_1 + \frac{\gamma}{\delta}, w_2 + \frac{\alpha}{\beta}\right)\right]\varepsilon + o(\varepsilon) \\ &= h_1(w_1,w_2)\varepsilon + o(\varepsilon)\end{aligned} \tag{3.54}$$

where Taylor series expansions of the f's in powers of ε were used. Similarly,

$$\begin{aligned}b_{22}w_2^2 - \varepsilon w_2 g_2(w_1,w_2) &= \frac{\beta}{\alpha}\, f_2\!\left(\frac{\gamma}{\delta}, \frac{\alpha}{\beta}\right)w_2^2 + f_2\!\left(\frac{\gamma}{\delta}, \frac{\alpha}{\beta}\right)w_2 \\ &\quad + f_{2x}\!\left(\frac{\gamma}{\delta}, \frac{\alpha}{\beta}\right)w_1w_2\end{aligned} \tag{3.55}$$

$$- w_2 f_2\left(w_1 + \frac{\gamma}{\delta}, w_2 + \frac{\alpha}{\beta}\right)\varepsilon + o(\varepsilon)$$
$$= h_2(w_1,w_2)\varepsilon + o(\varepsilon)$$

Hence if for some sufficiently large c, $h_i(w_1,w_2) < 0$, i = 1, 2, on Γ_c, then $\dot{V}(w_1,w_2) < 0$ on Γ_c and we would be in a situation of diminishing returns. Then, together with the previous increasing returns, by utilizing the Poincaré-Bendixon theorem, we could conclude the existence of a stable limit cycle. This is summarized in the following theorem.

THEOREM 3.1 *Let* $[\;]_1$ *be defined by* (3.41), $h_i(w_1,w_2)$, i = 1, 2, *be defined by* (3.54) *and* (3.55) *and* $\tilde{\Gamma}_c$ *be the translate of* Γ_c *into the* x-y *plane. Then if* $f_i(x,y)$, i = 1, 2, *are such that* $[\;]_1 > 0$, *and* $h_i(x - x^*, y - y^*) \leq 0$, i = 1, 2, *on* $\tilde{\Gamma}_c$ *for some sufficiently large* c, *and not both* h_1 *and* h_2 *simultaneously zero on* Γ_c, *then there exists a stable limit cycle of system* (3.29) *for sufficiently small* $\varepsilon > 0$.

We can note that the condition $[\;]_1 > 0$ can be replaced by $[\;]_i = 0$, $i = 1, \ldots, n - 1$, $[\;]_n > 0$.

It is now of interest to consider the case where

$$f_1\left(\frac{\gamma}{\delta}, \frac{\alpha}{\beta}\right) = f_2\left(\frac{\gamma}{\delta}, \frac{\alpha}{\beta}\right) = 0 \tag{3.56}$$

but

$$f_{1x}\left(\frac{\gamma}{\delta}, \frac{\alpha}{\beta}\right)^2 + f_{1y}\left(\frac{\gamma}{\delta}, \frac{\alpha}{\beta}\right)^2 + f_{2x}\left(\frac{\gamma}{\delta}, \frac{\alpha}{\beta}\right)^2 + f_{2y}\left(\frac{\gamma}{\delta}, \frac{\alpha}{\beta}\right)^2 = \kappa^2 + \lambda^2 + \mu^2 + \nu^2 \neq 0 \tag{3.57}$$

We refer to such a situation as a linear perturbation. Here the equilibrium is unperturbed, i.e.,

$$x^* = \frac{\gamma}{\delta}, \qquad y^* = \frac{\alpha}{\beta} \tag{3.58}$$

In this case several simplifications of the previous analysis occur. First, from (3.46),

$$b_{11} = -\varepsilon\kappa, \qquad b_{12} = -\frac{\beta\gamma}{\delta} - \varepsilon\lambda,$$
$$b_{21} = \frac{\alpha\delta}{\beta} - \varepsilon\mu, \qquad b_{22} = -\varepsilon\nu \tag{3.59}$$

Further, from (3.41) (or since $x_i^* = y_i^* = 0$, $i = 1, 2, \ldots$),

$$[\ \ ]_1 = -\kappa - \nu, \qquad [\ \ ]_2 = [\ \ ]_3 = \cdots = 0 \tag{3.60}$$

Hence if $\kappa + \nu < 0$, then the equilibrium $(\gamma/\delta, \alpha/\beta)$ is an unstable spiral point. Also

$$h_1(w_1,w_2) = \lambda w_1 w_2 - f_1\left(w_1 + \frac{\gamma}{\delta}, w_2 + \frac{\alpha}{\beta}\right)w_1$$
$$h_2(w_1,w_2) = \mu w_1 w_2 - f_2\left(w_1 + \frac{\gamma}{\delta}, w_2 + \frac{\alpha}{\beta}\right)w_2 \tag{3.61}$$

The statement of Theorem 3.1 continues to hold, but its practical aspects are somewhat simplified.

The previous analysis is taken from Freedman and Waltman (1975a). The analysis in the case where the perturbation is quadratic or higher is substantially different and will be presented in the next section.

3.7 QUADRATIC AND HIGHER PERTURBATIONS

In this section we again consider system (3.29), but under the conditions

$$f_1\left(\frac{\gamma}{\delta}, \frac{\alpha}{\beta}\right) = f_2\left(\frac{\gamma}{\delta}, \frac{\alpha}{\beta}\right) = f_{1x}\left(\frac{\gamma}{\delta}, \frac{\alpha}{\beta}\right) = f_{1y}\left(\frac{\gamma}{\delta}, \frac{\alpha}{\beta}\right)$$
$$= f_{2x}\left(\frac{\gamma}{\delta}, \frac{\alpha}{\beta}\right) = f_{2y}\left(\frac{\gamma}{\delta}, \frac{\alpha}{\beta}\right) = 0$$

In this case, since now $[\ \ ]_i = 0$, $i = 1, 2, \ldots$, there is no hope of using the previous analysis to conclude the existence of a periodic orbit. A different technique due to Loud (1959, 1964) is then needed. Before carrying out the required computations, we make certain transformations which simplify both the analysis and the equations. The following analysis is due to Freedman and Waltman (1975b).

We first let

$$\begin{aligned} w_1 &= x - \frac{\gamma}{\delta}, & w_2 &= y - \frac{\alpha}{\beta} \\ u &= \delta\sqrt{\alpha}\, w_1, & v &= \beta\sqrt{\gamma}\, w_2 \\ s &= \sqrt{\alpha\gamma}\, t \end{aligned} \tag{3.62}$$

Substituting into system (3.29) and simplifying gives

$$\begin{aligned} \dot{u} &= -v - auv - \varepsilon k_1(u,v) \\ \dot{v} &= u + buv - \varepsilon k_2(u,v) \end{aligned}, \qquad \left(\cdot = \frac{d}{ds}\right) \tag{3.63}$$

where

$$\begin{aligned} a &= \frac{1}{\gamma\sqrt{\alpha}}, \qquad b = \frac{1}{\alpha\sqrt{\gamma}} \\ k_1(u,v) &= \frac{\delta}{\sqrt{\gamma}}\, f_1\!\left(\frac{u}{\delta\sqrt{\alpha}} + \frac{\gamma}{\delta}, \frac{v}{\beta\sqrt{\gamma}} + \frac{\alpha}{\beta}\right) \\ k_2(u,v) &= \frac{\beta}{\sqrt{\alpha}}\, f_2\!\left(\frac{u}{\delta\sqrt{\alpha}} + \frac{\gamma}{\delta}, \frac{v}{\beta\sqrt{\gamma}} + \frac{\alpha}{\beta}\right) \end{aligned} \tag{3.64}$$

Further note that

$$k_i(0,0) = k_{iu}(0,0) = k_{iv}(0,0) = 0, \qquad i = 1,2 \tag{3.65}$$

It is system (3.63) that is examined for a periodic solution.

We first must investigate the variational equation of system (3.63). The unperturbed system

$$\begin{aligned} \dot{u} &= -v - auv \\ \dot{v} &= u + buv \end{aligned} \tag{3.66}$$

is, of course, equivalent to the Lotka-Volterra system, and so all solutions of (3.66) with initial conditions $u(0) > -1/b$, $v(0) > -1/a$, are periodic. Let $\phi_1(s)$, $\phi_2(s)$ be the solution of system (3.66) such that

$$\phi_1(0) = A > 0, \qquad \phi_2(0) = 0 \tag{3.67}$$

Then the variational equation corresponding to ϕ_1 and ϕ_2 is

$$\begin{aligned} \dot{z}_1 &= -a\phi_2(s)z_1 - [1 + a\phi_1(s)]z_2 \\ \dot{z}_2 &= [1 + b\phi_2(s)]z_1 + b\phi_1(s)z_2 \end{aligned} \tag{3.68}$$

As is well known, one solution of (3.68) is

$$p_1(s) = \dot{\phi}_1(s), \qquad p_2(s) = \dot{\phi}_2(s) \tag{3.69}$$

and, as is easily verified

$$\begin{aligned} p_1(0) &= 0, & p_2(0) &= A, \\ \dot{p}_1(0) &= -(1 + aA)A, & \dot{p}_2(0) &= bA^2 \end{aligned} \tag{3.70}$$

Now let $q_1(s)$, $q_2(s)$ be a second, linearly independent solution of system (3.68) with initial conditions

$$q_1(0) = \frac{1}{A}, \qquad q_2(0) = 0 \tag{3.71}$$

and hence $\dot{q}_1(0) = 0$, $\dot{q}_2(0) = 1/A$.

Let $u(s,\xi,\varepsilon)$, $v(s,\xi,\varepsilon)$ be that solution of (3.63) such that $u(0,\xi,\varepsilon) = A + \xi$, $v(0,\xi,\varepsilon) = 0$. We wish to determine ξ as a function of ε and τ as a function of ε so that

$$u(T + \tau, \xi, \varepsilon) = A + \xi, \qquad v(T + \tau, \xi, \varepsilon) = 0 \tag{3.72}$$

If we can find τ and ξ so that (3.72) is satisfied, we will have found a periodic solution, which is the object of the exercise. To accomplish this, we define the functions $F_i(\tau,\xi,\varepsilon)$, $i = 1, 2$, by

$$\begin{aligned} F_1(\tau,\xi,\varepsilon) &= u(T + \tau, \xi, \varepsilon) - A - \xi \\ F_2(\tau,\xi,\varepsilon) &= v(T + \tau, \xi, \varepsilon) \end{aligned} \tag{3.73}$$

Then satisfying (3.72) is equivalent to being able to solve $F_i(\tau,\xi,\varepsilon) = 0$, $i = 1, 2$, for τ and ξ as functions of ε for sufficiently small ε. We will do this by using the implicit function theorem. It will turn out that critical cases of the implicit function theorem are needed (see Sec. B.2 of the Appendix). To this end it will be necessary to evaluate $u(s,\xi,\varepsilon)$ and $v(s,\xi,\varepsilon)$ together with certain of their derivatives at $\xi = \varepsilon = 0$.

To begin with, it is clear that

$$
\begin{aligned}
&u(s,0,0) = \phi_1(s), \qquad v(s,0,0) = \phi_2(s) \\
&\dot{u}(s,0,0) = p_1(s), \qquad \dot{v}(s,0,0) = p_2(s) \\
&\ddot{u}(s,0,0) = -a\phi_2(s)p_1(s) - [1 + a\phi_1(s)]p_2(s) \\
&\ddot{v}(s,0,0) = [1 + b\phi_2(s)]p_1(s) + b\phi_1(s)p_2(s)
\end{aligned}
\tag{3.74}
$$

Next we compute the first partial derivatives of u and v with respect to ξ and ε. The terms $u_\xi(s,\xi,\varepsilon)$ and $v_\xi(s,\xi,\varepsilon)$ satisfy

$$
\begin{aligned}
&\dot{u}_\xi = -avu_\xi - (1 + au)v_\xi - \varepsilon k_{1u}(u,v)u_\xi - \varepsilon k_{1v}(u,v)v_\xi \\
&\dot{v}_\xi = (1 + bv)u_\xi + buv_\xi - \varepsilon k_{2u}(u,v)u_\xi - \varepsilon k_{2v}(u,v)v_\xi \\
&u_\xi(0,\xi,\varepsilon) = 1, \qquad v_\xi(0,\xi,\varepsilon) = 0
\end{aligned}
\tag{3.75}
$$

At $\xi = \varepsilon = 0$, $u_\xi(s,0,0)$ and $v_\xi(s,0,0)$ satisfy

$$
\begin{aligned}
&\dot{u}_\xi = -a\phi_2(s)u_\xi - [1 + a\phi_1(s)]v_\xi \\
&\dot{v}_\xi = [1 + b\phi_2(s)]u_\xi + b\phi_1(s)v_\xi \\
&u_\xi(0,0,0) = 1, \qquad v_\xi(0,0,0) = 0
\end{aligned}
\tag{3.76}
$$

Comparing this with the variational system (3.68), we find that

$$
\begin{aligned}
&u_\xi(s,0,0) = Aq_1(s), \qquad v_\xi(s,0,0) = Aq_2(s) \\
&\dot{u}_\xi(s,0,0) = -aA\phi_2(s)q_1(s) - A[1 + a\phi_1(s)]q_2(s) \\
&\dot{v}_\xi(s,0,0) = A[1 + b\phi_2(s)]q_1(s) + bA\phi_1(s)q_2(s)
\end{aligned}
\tag{3.77}
$$

Computing u_ε and v_ε, in a similar manner to the preceding, we get

$$
\begin{bmatrix} u_\varepsilon(s,0,0) \\ v_\varepsilon(s,0,0) \end{bmatrix} = -\Phi(s) \int_0^s \frac{1 + aA}{[1 + a\phi_1(r)][1 + b\phi_2(r)]}
\begin{bmatrix} p_2(r)k_1[\phi_1(r),\phi_2(r)] - p_1(r)k_2[\phi_1(r),\phi_2(r)] \\ -q_2(r)k_1[\phi_1(r),\phi_2(r)] + q_1(r)k_2[\phi_1(r),\phi_2(r)] \end{bmatrix} dr
\tag{3.78}
$$

It turns out that the second derivatives of u and v are also necessary. They are computed analagously to the first derivatives. In each case the equations satisfied by the second partial derivatives of u and v with respect to ξ and ε are written down, together with the initial conditions. Then ξ and ε are set equal to zero and known functions are substituted into the equations, which are then solved for the appropriate second derivatives. The details are lengthy, but routine, and so are omitted. The results are listed below.

$$\begin{bmatrix} u_{\xi\xi}(s,0,0) \\ v_{\xi\xi}(s,0,0) \end{bmatrix} = 2A^2(1 + aA)\Phi(s) \int_0^s \frac{q_1(r)q_2(r)}{[1 + a\phi_1(r)][1 + b\phi_2(r)]} \begin{bmatrix} -bp_1(r) - ap_2(r) \\ bq_1(r) + aq_2(r) \end{bmatrix} dr \tag{3.79}$$

$$\begin{bmatrix} u_{\xi\xi}(s,0,0) \\ v_{\xi\xi}(s,0,0) \end{bmatrix} = \Phi(s) \int_0^s \frac{A(1 + aA)}{[1 + a\phi_1(r)][1 + b\phi_2(r)]} \begin{bmatrix} p_2(r)r_1(r) - p_1(r)r_2(r) \\ -q_2(r)r_1(r) + q_1(r)r_2(r) \end{bmatrix} dr \tag{3.80}$$

where

$$\begin{aligned} r_1(s) &= -aq_1(s)v_\varepsilon(s,0,0) - aq_2(s)u_\varepsilon(s,0,0) \\ &\quad - k_{1u}[\phi_1(s),\phi_2(s)]q_1(s) - k_{1v}[\phi_1(s),\phi_2(s)]q_2(s) \\ r_2(s) &= bq_1(s)v_\varepsilon(s,0,0) + bq_2(s)u_\varepsilon(s,0,0) \\ &\quad - k_{2u}[\phi_1(s),\phi_2(s)] - k_{2v}[\phi_1(s),\phi_2(s)]q_2(s) \end{aligned} \tag{3.81}$$

$$\begin{bmatrix} u_{\varepsilon\varepsilon}(s,0,0) \\ v_{\varepsilon\varepsilon}(s,0,0) \end{bmatrix} = \Phi(s) \int_0^s \frac{1 + aA}{[1 + a\phi_1(r)][1 + b\phi_2(r)]} \begin{bmatrix} p_2(r)\rho_1(r) - p_1(r)\rho_2(r) \\ -q_2(r)\rho_1(r) + q_1(r)\rho_2(r) \end{bmatrix} dr \tag{3.82}$$

where

$$\rho_1(s) = -2au_\varepsilon(s,0,0)v_\varepsilon(s,0,0) - 2k_{1u}[\phi_1(s),\phi_2(s)]u_\varepsilon(s,0,0) - 2k_{1v}[\phi_1(s),\phi_2(s)]v_\varepsilon(s,0,0) \tag{3.83}$$

$$\rho_2(s) = 2bu_\varepsilon(s,0,0)v_\varepsilon(s,0,0) - 2k_{2u}[\phi_1(s),\phi_2(s)]u_\varepsilon(s,0,0) - 2k_{2v}[\phi_1(s),\phi_2(s)]v_\varepsilon(s,0,0)$$

To utilize the implicit function theorem, we shall need to evaluate the F_i, i = 1, 2 [as given by (3.73)], and their first and second partial derivatives at $\tau = \xi = \varepsilon = 0$. This can be done in terms of the values of u and v and their partial derivatives as just computed, evaluated at s = T, the period. For details of these evaluations see Freedman and Waltman (1975b). We comment that $q_2(T)$ has been shown by computer to be in general negative, (see Table 3.1). The results follow.

$$F_i(0,0,0) = 0, \qquad i = 1, 2 \tag{3.84}$$

$$F_{1\tau}(0,0,0) = 0, \qquad F_{2\tau}(0,0,0) = A \tag{3.85}$$

$$F_{1\xi}(0,0,0) = 0, \qquad F_{2\xi}(0,0,0) = Aq_2(T) \tag{3.86}$$

TABLE 3.1

a	b	A	T	$q_2(T)$	$\int_0^T q_2(s)\,ds$
2.0	1.0	1.0	6.8848	-0.9047	0.00015
1.0	2.0	1.0	7.1074	-1.3580	0.00049
3.0	1.0	1.0	7.2461	-1.3573	0.00094
1.0	3.0	1.0	7.9355	-2.6427	0.00219
1.0	2.0	2.0	8.7441	-1.7785	0.00023
1.0	1.0	1.0	6.6094	-0.5358	0.00048

$$F_{1\varepsilon}(0,0,0) = \frac{1 + aA}{A} \int_0^T \left[\frac{-p_2(r)k_1[\phi_1(r),\phi_2(r)]}{[1 + a\phi_1(r)][1 + b\phi_2(r)]} + \frac{p_1(r)k_2[\phi_1(r),\phi_2(r)]}{[1 + a\phi_1(r)][1 + b\phi_2(r)]} \right] dr \tag{3.87}$$

$$F_{2\varepsilon}(0,0,0) = (1 + aA) \int_0^T \left[\frac{[-q_2(T)p_2(r) + Aq_2(r)]k_1[\phi_1(r),\phi_2(r)]}{[1 + a\phi_1(r)][1 + b\phi_2(r)]} + \frac{[q_2(T)p_1(r) - Aq_1(r)]k_2[\phi_1(r),\phi_2(r)]}{[1 + a\phi_1(r)][1 + b\phi_2(r)]} \right] dr$$

$$F_{1\tau\tau}(0,0,0) = -(1 + aA)A, \qquad F_{2\tau\tau}(0,0,0) = bA^2 \tag{3.88}$$

$$\begin{aligned} F_{1\tau\xi}(0,0,0) &= -(1 + aA)Aq_2(T), \\ F_{2\tau\xi}(0,0,0) &= 1 + bA^2q_2(T) \end{aligned} \tag{3.89}$$

$$\begin{aligned} F_{1\tau\varepsilon}(0,0,0) &= -(1 + aA)F_{2\varepsilon}(0,0,0) - k_1(A,0) \\ F_{2\tau\varepsilon}(0,0,0) &= F_{1\varepsilon}(0,0,0) + bAF_{2\varepsilon}(0,0,0) - k_2(A,0) \end{aligned} \tag{3.90}$$

$$F_{1\xi\xi}(0,0,0) = -2A(1 + aA) \int_0^T \frac{q_1(r)q_2(r)[bp_1(r) + ap_2(r)]}{[1 + a\phi_1(r)][1 + b\phi_2(r)]} dr$$

$$F_{2\xi\xi}(0,0,0) = 2A^2(1 + aA) \int_0^T \left[\frac{q_1(r)q_2(r)\{b[-q_2(T)p_1(r) + Aq_1(r)]}{[1 + a\phi_1(r)][1 + b\phi_2(r)]} + \frac{a[-q_2(T)p_2(r) + Aq_2(r)]\}}{[1 + a\phi_1(r)][1 + b\phi_2(r)]} \right] dr \tag{3.91}$$

$$F_{1\xi\varepsilon}(0,0,0) = (1 + aA) \int_0^T \frac{p_2(r)r_1(r) - p_1(r)r_2(r)}{[1 + a\phi_1(r)][1 + b\phi_2(r)]} dr \tag{3.92}$$

$$F_{2\xi\varepsilon}(0,0,0) = A(1 + aA) \int_0^T \left[\frac{[q_2(T)p_2(r) - Aq_2(r)]r_1(r)}{[1 + a\phi_1(r)][1 + b\phi_2(r)]} + \frac{[-q_2(T)p_1(r) + Aq_1(r)]r_2(r)}{[1 + a\phi_1(r)][1 + b\phi_2(r)]} \right] dr$$

$$F_{1\varepsilon\varepsilon}(0,0,0) = \frac{1 + aA}{A} \int_0^T \frac{p_2(r)\rho_1(r) - p_1(r)\rho_2(r)}{[1 + a\phi_1(r)][1 + b\phi_2(r)]} dr$$

$$F_{2\varepsilon\varepsilon}(0,0,0) = (1 + aA) \int_0^T \left[\frac{[q_2(T)p_2(r) - Aq_2(r)]\rho_1(r)}{[1 + a\phi_1(r)][1 + b\phi_2(r)]} + \frac{[-q_2(T)p_1(r) + Aq_1(r)]\rho_2(r)}{[1 + a\phi_1(r)][1 + b\phi_2(r)]}\right] dr \tag{3.93}$$

We now consider the first of formulas (3.91) and simplify $F_{1\xi\xi}(0,0,0)$ even further. First we note that

$$\frac{bp_1(s) + ap_2(s)}{[1 + a\phi_1(s)][1 + b\phi_2(s)]} = \frac{1}{1 + b\phi_2(s)} - \frac{1}{1 + a\phi_1(s)}$$

Further,

$$\int_0^s \frac{q_1(r)q_2(r)}{[1 + a\phi_1(r)]} dr = -\int_0^s \frac{q_1(r)}{1 + a\phi_1(r)} \frac{d}{dr}\left[\frac{q_1(r)}{1 + a\phi_1(r)}\right] dr$$

$$= -\frac{q_1(s)^2}{2[1 + a\phi_1(s)]^2} + \frac{1}{2A^2(1 + aA)^2}$$

and similarly,

$$\int_0^s \frac{q_1(r)q_2(r)}{1 + b\phi_2(r)} dr = \frac{q_2(s)^2}{2\left[1 + b\phi_2(s)\right]^2}$$

Hence from (3.91) we get

$$F_{1\xi\xi}(0,0,0) = -A(1 + aA)q_2(T)^2 \tag{3.94}$$

We can finally investigate the conditions which lead to periodic solutions of system (3.63) for sufficiently small positive ε. As we have said, if $F_i(\tau,\xi,\varepsilon) = 0$, $i = 1,2$, can be solved for τ and ξ as functions of ε, we have found the required periodic solution. Now by (3.85), since $F_{2\tau}(0,0,0) = A \neq 0$, $F_2(\tau,\xi,\varepsilon) = 0$ can be solved for τ as a function of ξ and ε, $\tau = \tau(\xi,\varepsilon)$, such that by (3.86) and (3.87),

$$\tau = -q_2(T)\xi - \frac{F_{2\varepsilon}(0,0,0)}{A}\varepsilon + \text{Higher Order Terms} \tag{3.95}$$

We now substitute this solution in $F_1(\tau,\xi,\varepsilon)$ and let

$$G(\xi,\varepsilon) = F_1[\tau(\xi,\varepsilon),\xi,\varepsilon] \tag{3.96}$$

Then

$$G(0,0) = F_1(0,0,0) = 0 \tag{3.97}$$

and

$$G_\xi(0,0) = F_{1\tau}(0,0,0)\frac{\partial\tau}{\partial\xi}(0,0) + F_{1\xi}(0,0,0) = 0 \tag{3.98}$$

by (3.85) and (3.86). Further,

$$\begin{aligned} G_\xi(0,0) &= F_{1\tau}(0,0,0)\frac{\partial\tau}{\partial\varepsilon}(0,0) + F_{1\varepsilon}(0,0,0) \\ &= F_{1\varepsilon}(0,0,0) \end{aligned} \tag{3.99}$$

by (3.85). It is now assumed, in order to apply critical cases of the implicit function theorem using no higher than quadratic terms, that k_1 and k_2 are so chosen that

$$F_{1\varepsilon}(0,0,0) = 0 \tag{3.100}$$

The reason for this is that if $G_\varepsilon(0,0) \neq 0$, then the implicit function theorem would demand that $G_{\xi\xi}(0,0) \neq 0$. But

$$\begin{aligned} G_{\xi\xi}(0,0) &= F_{1\tau\tau}(0,0,0)\frac{\partial\tau}{\partial\xi}(0,0)^2 + 2F_{1\tau\xi}(0,0,0)\frac{\partial\tau}{\partial\xi}(0,0) \\ &\quad + F_{1\tau}(0,0,0)\frac{\partial^2\tau}{\partial\xi^2}(0,0) + F_{1\xi\xi}(0,0,0) \\ &= -(1 + aA)Aq_2(T)^2 + 2(1 + aA)Aq_2(T)^2 \\ &\quad - A(1 + aA)q_2(T)^2 = 0 \end{aligned}$$

It may be that k_1 and k_2 are such that (3.100) holds for all $A > 0$. In that case we would expect the perturbed system to give a neutrally stable equilibrium (if the following hypotheses, which guarantee periodic solutions, also hold). However, if (3.100) holds only for certain discrete $A > 0$, this could give rise to a finite number (and possibly a unique) limit cycle.

Under the assumption that (3.100) holds, let

$$\xi = \eta\varepsilon \tag{3.101}$$

and define

$$H(\eta,\varepsilon) = \varepsilon^{-2}G(\eta\varepsilon,\varepsilon)$$

Computing $H(\eta,0)$ in a straightforward manner, you get

$$H(\eta,0) = G_{\xi\varepsilon}(0,0)\eta + \frac{1}{2}G_{\varepsilon\varepsilon}(0,0) \tag{3.102}$$

where

$$\begin{aligned} G_{\xi\varepsilon}(0,0) &= g_1(A,0)q_2(T)A^{-1} + (1 + aA)q_2(T)F_{2\varepsilon}(0,0,0) \\ &\quad + F_{1\xi\varepsilon}(0,0,0) \\ G_{\varepsilon\varepsilon}(0,0) &= (1 + aA)F_{2\varepsilon}(0,0,0)^2A^{-1} + g_1(A,0)F_{2\varepsilon}(0,0,0)A^{-1} \\ &\quad + F_{1\varepsilon\varepsilon}(0,0,0) \end{aligned} \tag{3.103}$$

where $F_{2\varepsilon}$, $F_{1\xi\varepsilon}$, and $F_{1\varepsilon\varepsilon}$ are given by formulas (3.87), (3.92), and (3.93), respectively.

The condition, then, that $H(\eta,0)$ have a simple root η_o is that

$$G_{\xi\varepsilon}(0,0) \neq 0 \tag{3.104}$$

and in this case

$$\eta_o = -\frac{G_{\varepsilon\varepsilon}(0,0)}{G_{\xi\varepsilon}(0,0)} \tag{3.105}$$

If (3.104) holds, then by the implicit function theorem $H(\eta,\varepsilon) = 0$ can be solved for η as a function of ε, and then, by (3.101), ξ may be found as a function of ε, and then τ as a function of ε; thus the required periodic solution is obtained. The preceding work is summarized in the following theorem.

THEOREM 3.2 *Let* $k_i(u,v) \in C^2$ *and* a, b, A, $k_i(u,v)$ *be such that* (3.100) *and* (3.104) *hold. Then system* (3.63) *has a periodic solution for sufficiently small* ε *(positive and negative) of period* $T + \tau(\varepsilon)$ *and initial values* $[A + \xi(\varepsilon), 0]$ *where* $\xi(\varepsilon) = \eta_o \varepsilon + o(\varepsilon)$ *and* $\tau(\varepsilon) = -[q_2(T)\eta_o + F_{2\varepsilon}(0,0,0)A^{-1}]\varepsilon + o(\varepsilon)$, *and where* η_o *is given by* (3.105).

For biological purposes it would be desirable that the periodic solution just found be stable. The stability of this solution will be investigated in the next section.

3.8 STABILITY OF THE PERTURBED PERIODIC SOLUTION

Here we shall investigate the stability of the periodic solution found in the preceding section. This will be done by examining the characteristic multipliers of the variational equation corresponding to this solution of system (3.63) (see Sec. D.5 of the Appendix).

Let u(s), v(s) be the periodic solution under discussion. Then the variational equation corresponding to this solution is

$$\begin{aligned} \dot{Z}_1 &= \{-av(s) - \varepsilon k_{1u}[u(s),v(s)]\}Z_1 \\ &\quad + \{-1 - au(s) - \varepsilon k_{1v}[u(s),v(s)]\}Z_2 \\ \dot{Z}_2 &= \{1 + bv(s) - \varepsilon k_{2u}[u(s),v(s)]\}Z_1 \\ &\quad + \{bu(s) - \varepsilon k_{2v}[u(s),v(s)]\}Z_2 \end{aligned} \tag{3.106}$$

Let $\ell_i(s)$ and $m_i(s)$, i = 1, 2, be those solutions of (3.106) such that

$$\ell_1(0) = m_2(0) = 1, \qquad \ell_2(0) = m_1(0) = 0 \tag{3.107}$$

Further, let

$$\Psi(s) = \begin{bmatrix} \ell_1(s) & m_1(s) \\ \ell_2(s) & m_2(s) \end{bmatrix} \tag{3.108}$$

Then the stability of the periodic solution is given by the eigenvalues of $\Psi(T)$. At $\varepsilon = 0$, system (3.106) reduces to system (3.68), and in that case both eigenvalues are 1. We wish to examine the eigenvalues for small positive ε.

The periodic solutions u(s) and v(s) may be expanded in powers of ε,

$$\begin{aligned} u(s) &= \phi_1(s) + u_1(s)\varepsilon + o(\varepsilon) \\ v(s) &= \phi_2(s) + v_1(s)\varepsilon + o(\varepsilon) \end{aligned} \tag{3.109}$$

We compute now $u_1(s)$ and $v_1(s)$ by differentiating $u[s,\xi(\varepsilon),\varepsilon]$ and $v[s,\xi(\varepsilon),\varepsilon]$ with respect to ε and evaluating at $\varepsilon = 0$. This gives

$$\begin{aligned} u_1(s) &= A\eta_o q_1(s) + u_\varepsilon(s,0,0) \\ v_1(s) &= A\eta_o q_2(s) + v_\varepsilon(s,0,0) \end{aligned} \tag{3.110}$$

where η_o is given by (3.105) and u_ε, v_ε by (3.78).

We now seek solutions of system (3.106) of the form

$$Z_i(s) = Z_{i0}(s) + \varepsilon Z_{i1}(s) + o(\varepsilon) \tag{3.111}$$

Substituting (3.109) and (3.111) into (3.106) and comparing coefficients gives the following systems for Z_{i0} and Z_{i1},

$$\begin{aligned} \dot{Z}_{10} &= -a\phi_2(s)Z_{10} + [-1 - a\phi_1(s)]Z_{20} \\ \dot{Z}_{20} &= [1 + b\phi_2(s)]Z_{10} + b\phi_1(s)Z_{20} \end{aligned} \tag{3.112}$$

$$\begin{aligned} \dot{Z}_{11} &= -a\phi_2(s)Z_{11} + [-1 - a\phi_1(s)]Z_{21} \\ &\quad + \{-av_1(s) - k_{1u}[\phi_1(s),\phi_2(s)]\}Z_{10}(s) \\ &\quad + \{-au_1(s) - k_{1v}[\phi_1(s),\phi_2(s)]\}Z_{20}(s) \\ \dot{Z}_{21} &= [1 + b\phi_2(s)]Z_{11} + b\phi_1(s)Z_{21} \\ &\quad + \{bv_1(s) - k_{2u}[\phi_1(s),\phi_2(s)]\}Z_{10}(s) \\ &\quad + \{bu_1(s) - k_{2v}[\phi_1(s),\phi_2(s)]\}Z_{20}(s) \end{aligned} \tag{3.113}$$

We now let

$$\begin{aligned} \ell_i(s) &= \ell_{i0}(s) + \varepsilon\ell_{i1}(s) + o(\varepsilon) \\ m_i(s) &= m_{i0}(s) + \varepsilon m_{i1}(s) + o(\varepsilon) \end{aligned} \tag{3.114}$$

Since (3.107) must be satisfied for all sufficiently small $\varepsilon \geq 0$, it must be that

$$\begin{aligned} &\ell_{10}(0) = m_{20}(0) = 1 \\ &\ell_{20}(0) = m_{10}(0) = \ell_{i1}(0) = m_{i1}(0) = 0, \qquad i = 1,\ 2 \end{aligned} \tag{3.115}$$

Hence from (3.112) we get immediately that

$$\begin{aligned} &\ell_{i0}(s) = A^{-1}q_i(s), \qquad m_{i0}(s) = Ap_i(s), \\ &i = 1,\ 2 \end{aligned} \tag{3.116}$$

It is now convenient to define $K_i(s)$, $i = 1, 2, 3$, by

$$K_1(s) = \begin{bmatrix} -q_2(s)[ap_2(s) + bp_1(s)] & -q_1(s)[ap_2(s) + bp_1(s)] \\ q_2(s)[aq_2(s) + bq_1(s)] & q_1(s)[aq_2(s) + bq_1(s)] \end{bmatrix} \tag{3.117}$$

$$K_2(s) = \begin{bmatrix} -v_\varepsilon(s,0,0)[ap_2(s) + bp_1(s)] & -u_\varepsilon(s,0,0)[ap_2(s) + bp_1(s)] \\ v_\varepsilon(s,0,0)[aq_2(s) + bq_1(s)] & u_\varepsilon(s,0,0)[aq_2(s) + bq_1(s)] \end{bmatrix} \tag{3.118}$$

$$K_3(s) = \begin{bmatrix} -p_2(s)k_{1u}[\phi_1(s),\phi_2(s)] & -p_2(s)k_{1v}[\phi_1(s),\phi_2(s)] \\ \quad - p_1(s)k_{2u}[\phi_1(s),\phi_2(s)] & \quad - p_1(s)k_{2v}[\phi_1(s),\phi_2(s)] \\ q_2(s)k_{1u}[\phi_1(s),\phi_2(s)] & q_2(s)k_{1v}[\phi_1(s),\phi_2(s)] \\ \quad + q_1(s)k_{2u}[\phi_1(s),\phi_2(s)] & \quad + q_2(s)k_{2v}[\phi_1(s),\phi_2(s)] \end{bmatrix} \tag{3.119}$$

and $K(s)$ by

$$K(s) = \frac{K_1(s) + K_2(s) + K_3(s)}{[1 + a\phi_1(s)][1 + b\phi_2(s)]} \tag{3.120}$$

Then from (3.113), utilizing (3.110), (3.114), and (3.115), we find that

$$\begin{bmatrix} \ell_{11}(s) \\ \ell_{21}(s) \end{bmatrix} = \frac{1 + aA}{A}\,\Phi(s) \int_0^s K(r) \begin{bmatrix} q_1(r) \\ q_2(r) \end{bmatrix} dr \tag{3.121}$$

and

$$\begin{bmatrix} m_{11}(s) \\ m_{21}(s) \end{bmatrix} = (1 + aA)A\Phi(s) \int_0^s K(r) \begin{bmatrix} p_1(r) \\ p_2(r) \end{bmatrix} dr \tag{3.122}$$

We now can compute the eigenvalues of $\Psi(T)$

$$\Psi(T) = \begin{bmatrix} 1 + \varepsilon\ell_{11}(T) + o(\varepsilon) & \varepsilon m_{11}(T) + o(\varepsilon) \\ Aq_2(T) + \varepsilon\ell_{21}(T) + 0(\varepsilon) & 1 + \varepsilon m_{21}(T) + o(\varepsilon) \end{bmatrix} \tag{3.123}$$

where $\ell_{i1}(T)$ and $m_{i1}(T)$, $i = 1, 2$ are obtained from (3.121) and (3.122), respectively, by setting $s = T$. From this the eigenvalues are seen to be of the form

$$1 + \varepsilon \frac{[\ell_{11}(T) + m_{21}(T)]}{2} + o(\varepsilon) \pm \sqrt{Aq_2(T)m_{11}(T)\varepsilon + o(\varepsilon)} \tag{3.124}$$

Using the notation of Loud (1959, 1964), the preceding proves the following theorem, where ε is assumed to be sufficiently small and positive.

THEOREM 3.3 *(i) If* $\ell_{11}(T) + m_{21}(T) < 0$ *and* $q_2(T)m_{11}(T) < 0$, *the periodic solution is completely stable. (ii) If* $q_2(T)m_{11}(T) < 0$ *and* $\ell_{11}(T) + m_{21}(T) > 0$, *the periodic solution is completely unstable. (iii) If* $q_2(T)m_{11}(T) > 0$, *the periodic solution is directly unstable.*

Under part (i), solutions beginning near the periodic solution will approach it closer as time goes on. This is a highly desirable situation biologically.

NOTES ON THE LITERATURE

The material in the last three sections may be found in Freedman and Waltman (1975a,b,c).

Certain papers have been written on topics related to Lotka-Volterra predator-prey systems. Models in which one or more of the variables is substituted by its logarithm (thus making the specific growth rate sublinear) are found in Gomatam (1974), Hassell and Huffaker (1969), Strickfaden and Lawrence (1975), whereas a model considering a predator-food situation is in Ishihara et al. (1972). Leslie (1957) has indicated that Gause's data fit a Lotka-Volterra model with carrying capacity. Frame (1974) and Grasman and Veling (1973) have concerned themselves with the period of solutions of the Lotka-Volterra system. Smith and Mead (1974) have considered a Lotka-Volterra model with age structure. Vandemeer (1973a) has set up a model of migration similar to the Lotka-Volterra model. Alekseev and Svetlosanov (1974), Levin (1974), and Walter (1974) consider models with dispersion or dissipation. Trubatch and Franco (1974) have computed Lagrangians for the Lotka-Volterra system (and others), Comins and Blatt (1974) have utilized diffusion terms to write the Lotka-Volterra system as a partial differential equation. Dutt et al. (1975) have also applied perturbation theory to Lotka-Volterra systems. Cushing (1976c) has considered Lotka-Volterra systems with diminishing returns, where the coefficients are periodic or almost periodic functions of time.

For papers involving Lotka-Volterra-type systems with time delays and/or integral terms (learning effects) see Bownds and Cushing (1975), Cushing (1976a), Goh et al. (1974), May (1973d), Ross (1972). For general papers on Lotka-Volterra systems and/or papers on statistical aspects, see Bartlett (1957), Canale (1970), Gilpin (1974b), Goel et al (1971), and Kilmer (1972).

EXERCISES

3.1 Let $x(t)$ and $y(t)$ denote solutions to the Lotka-Volterra system (3.1), and let $\langle Z(t)\rangle$ denote the time average of a function $Z(t)$, i.e., $\langle Z(t)\rangle = 1/T \int_0^T Z(t)\,dt$, where T is the period of $x(t)$ and $y(t)$.

(a) Compute $\langle x(t)\rangle$, $\langle y(t)\rangle$, $\langle x(t)y(t)\rangle$, $\langle x'(t)\rangle$, $\langle y'(t)\rangle$, $\langle x^2(t)\rangle$, $\langle y^2(t)\rangle$, $\langle x^3(t)\rangle$, $\langle x^2(t)y(t)\rangle$, $\langle x(t)y^2(t)\rangle$, $\langle y^3(t)\rangle$.

(b) Can a technique be found to compute $\langle x^p(t) \rangle$ in terms of lower-order time averages?

3.2 For system (3.17) analyze the case of mixed returns, i.e., $c_1 c_2 < 0$.

3.3 Consider the modified model (3.28) suggested by Samuelson.

(a) For ε small, analyze the model by perturbation techniques.

(b) (Open problem) For ε not necessarily small, classify the functions R_1 and R_2 which yield a *unique* limit cycle for this model.

3.4 (Open problem) Analyze a general time-dependent perturbed system for periodic solutions.

(a) Such a system may be of the form

$$x' = x(\alpha - \beta y) - \varepsilon f_1(t,x,y)$$

$$y' = y(-\gamma + \delta x) - \varepsilon f_2(t,x,y)$$

where the f_i's are periodic in t of period ω.

(b) An even more general system would be of the form

$$x' = x(\alpha - \beta y) - f_1(t,x,y,\varepsilon)$$

$$y' = y(-\gamma + \delta x) - f_2(t,x,y,\varepsilon)$$

such that the f_i's are periodic in t, and $f_i(t,x,y,0)$ is strictly a function of t (which may be identically zero).

3.5 (Open problem) For the Lotka-Volterra system, can one find a general expression for the period in terms of the given parameters? (See Grasman and Veling, 1973.)

Chapter 4

INTERMEDIATE PREDATOR-PREY MODELS

A: CONTINUOUS MODELS

In the first part of this chapter a continuous model, which is a generalization of a model due to Gause (1934), will be described and analyzed. In particular, the equilibria will be studied for their stability properties and the solutions analyzed in the phase plane. A modification of this model to include a source term for the prey is also included.

A perturbed form of this model will be used as a model for pest control in Chap. 6.

4.1 A GENERALIZED GAUSE MODEL

In Gause (1934) and Gause et al. (1936) the following model was considered as a model for predator-prey interactions

$$\begin{aligned} x' &= \alpha x - yp(x) \\ y' &= y[-\gamma + cp(x)] \end{aligned} \tag{4.1}$$

In part A of this chapter we analyze a slightly more general form of this model, as an intermediate model of predator-prey interactions. This model takes the form

$$\begin{aligned} x' &= xg(x) - yp(x) \\ y' &= y[-\gamma + q(x)] \end{aligned} \tag{4.2}$$

Here $g(x)$ is the specific growth rate of the prey in the absence of any predators and $p(x)$ is the predator response function for the predator with respect to that particular prey.

The properties of $g(x)$ were discussed in Chap. 1. The assumptions on $g(x)$ here will be the same, so that the single-species growth of Chap. 1 will apply to system (4.2) in the absence of any predators. For completeness we reiterate the assumptions on $g(x)$ here.

$$g(0) = \alpha > 0, \quad g(x) \text{ is continuous and differentiable} \quad \text{for } x \geq 0, \quad g_x(x) \leq 0 \tag{4.3}$$

When the environment has a carrying capacity, we have

$$\exists K > 0 \ni g(K) = 0 \tag{4.4}$$

This last assumption is, of course, biologically realistic.

The term $p(x)$ will have the properties described in Chap. 2, i.e., we assume

$$p(0) = 0, \quad p(x) \text{ is continuous and differentiable} \quad \text{for } x \geq 0, \quad p_x(x) > 0 \tag{4.5}$$

As a consequence, we have

$$\lim_{x\to\infty} p(x) = p_\infty, \qquad 0 < p_\infty \leq \infty \tag{4.6}$$

For definiteness we let

$$p_x(0) = \beta > 0 \tag{4.7}$$

The first part of (4.2) then states that the prey growth is enhanced by its own presence in a manner analogous to single-species growth and is diminished by an amount proportional to the number of predators present multiplied by the predator response to the prey.

In Gause's model (4.1), $q(x) = cp(x)$. It will be helpful to think of $q(x)$ in this manner. Essentially, $q(x)$ will have properties similar to $p(x)$, namely,

$$q(0) = 0, \quad q(x) \text{ is continuous and differentiable} \quad \text{for } x \geq 0, \quad q_x(x) > 0 \tag{4.8}$$

$$\lim_{x\to\infty} q(x) = q \ , \qquad 0 < q_\infty \leq \infty \tag{4.9}$$

and

$$q_x(0) = \delta > 0 \tag{4.10}$$

The second statement of (4.2) now describes the growth of the predator population. In the absence of prey the predator population declines. The growth is enhanced by the presence of prey by an amount proportional to the number of prey. In some sense, $q(x)$ may be interpreted as spelling out what proportion of prey eaten becomes predator. This proportionality may vary with prey density. If not, then indeed $q(x) = cp(x)$.

In the following sections, we shall analyze the equilibria and phase motions of system (4.2).

Note that throughout this chapter by "stable" we mean asymptotically stable in the sense of Liapunov.

4.2 EXISTENCE OF EQUILIBRIA

By (4.5), system (4.2) always has at least one equilibrium, namely, (0,0). We will certainly want to consider the case where (4.4) holds, especially for applications to pest control theory in Chap. 6. In this case there will be a second equilibrium on the axes at (K,0).

The equilibrium of greatest interest would be an equilibrium interior to the first quadrant, so we seek conditions for such an equilibrium to exist.

The first condition is that

$$q_\infty > \gamma \tag{4.11}$$

Biologically, this is necessary for the persistence of the ecosystem, for if $q_\infty \leq \gamma$, then $-\gamma + q(x) < 0$ for all $x > 0$ and $y' < 0$. Hence $\lim_{t\to\infty} y(t) = y_0 \geq 0$, and $\lim_{t\to\infty} y'(t) = 0$. But if $y_0 > 0$, by the second equation of (4.2), $y'(\infty) < 0$ and so $y_0 = 0$. This implies that the predator species goes extinct. Another way of looking at it is that $q_\infty \leq \gamma$ means that no matter how densely packed and available the prey are, the predator's predation is so poor that its growth cannot be sustained. Whereas this may be true for a system consisting of chipmunks trying to live by eating elephants, in real-life situations, for models simulating predator-prey systems where both species persist, (4.11) is valid.

Suppose, then, that (4.11) holds. Then by (4.5) there is a unique $x^* > 0$ such that

$$q(x^*) = \gamma \tag{4.12}$$

(see Fig. 4.1b). x^* is the x value of the potential interior equilibrium.

To determine the y value of the equilibrium we merely solve $xg(x) - yp(x) = 0$ for y at x^* and get

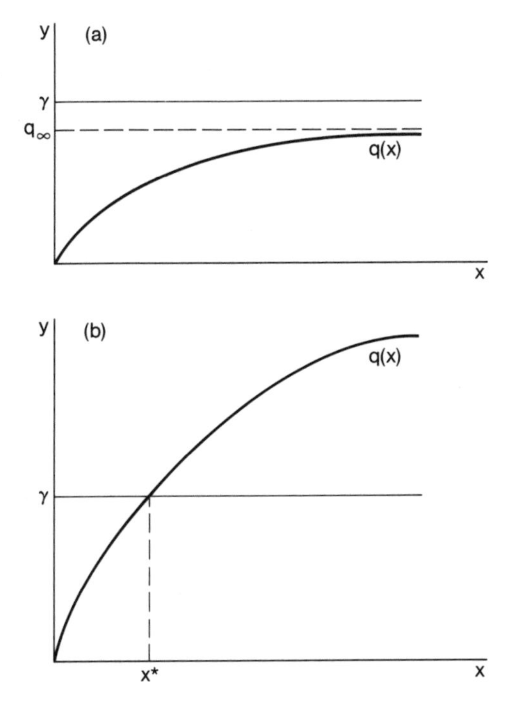

Fig. 4.1 (a) $q_\infty < \gamma$, predator goes extinct. (b) $q_\infty > \gamma$.

$$y^* = \frac{x^*g(x^*)}{p(x^*)} \tag{4.13}$$

To ensure that $y^* > 0$, however, we must make a second assumption, which again is biologically proper. This assumption is made in the case the environment has a carrying capacity, namely,

$$x^* < K \tag{4.14}$$

Then $g(x^*) > 0$.

Biologically, if $x^* \geq K$, then the predators again go extinct. This is clear, for even in the absence of predators the prey population is bounded above by K or brought down to K if initially above it. With predators present, the population will be brought below K and stay below. Then if $x^* \geq K$, as before, $-\gamma + q(x^*) < 0$ and $y' < 0$, causing the predators to go extinct. Then (x^*,y^*) is the desired interior equilibrium.

In the next section the stabilities of these equilibria are investigated.

4.3 STABILITY OF THE EQUILIBRIA

Here we are concerned with the stability of the three equilibrium points, the existence of which was discussed in the previous section. It will be assumed throughout this section that the conditions guaranteeing their existence are fulfilled. In order they are

(i) $(0,0)$
(ii) $(K,0)$
(iii) (x^*,y^*)

From Sec. D of the appendix we see that it may be possible to analyze the stability by computing the real parts of the eigenvalues of the Jacobian matrix evaluated at these equilibria. With this in mind we now compute the Jacobian of system (4.2). It is easily seen to be

$$J(x,y) = \begin{bmatrix} xg_x(x) + g(x) - yp_x(x) & -p(x) \\ yq_x(x) & -\gamma + q(x) \end{bmatrix} \tag{4.15}$$

For the equilibrium (i), J reduces to

$$J(0,0) = \begin{bmatrix} \alpha & 0 \\ 0 & -\gamma \end{bmatrix} \tag{4.16}$$

whose eigenvalues are α and $-\gamma$. Hence (0,0) is a hyperbolic or saddle point. It is clear from the second statement of (4.2) that on the y-axis the flow is toward the origin (i.e., in the absence of prey, the predators become extinct) and from the first of these equations that on the x-axis for small x, the flow is away from the origin [i.e. for small numbers of prey and predators, the prey population increases (see Fig. 4.2)]. This last consequence is taken as an axiom in the Kolmogorov model (Chap. 5).

Consider now equilibrium (ii). Here

$$J(K,0) = \begin{bmatrix} Kg_x(K) & -p(K) \\ 0 & -\gamma + q(K) \end{bmatrix} \tag{4.17}$$

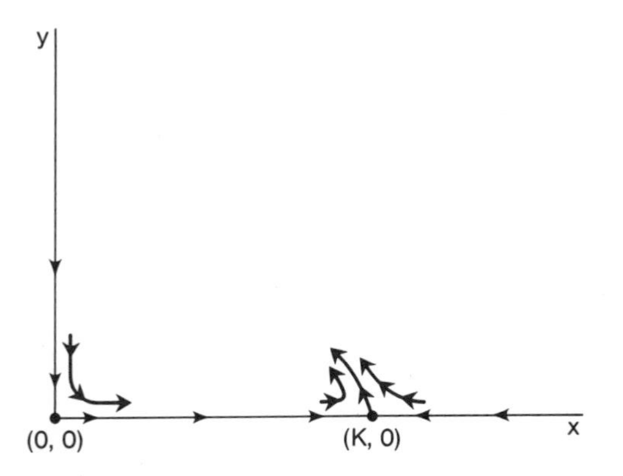

Fig. 4.2 Flow of solutions of system (4.2) along the axes and near equilibria (i) and (ii).

At this time we are going to assume that inequality in (4.3) holds at $x = K$, i.e.,

$$g_x(K) < 0 \tag{4.18}$$

Then, by (4.14), $-\gamma + q(K) > 0$ and once again, since the eigenvalues of $J(K,0)$ are $Kg_x(K)$ and $-\gamma + q(K)$, $(K,0)$ is a hyperbolic point. Clearly, the flow near $(K,0)$ on the x-axis is toward $x = K$. Then there is a direction oblique to the x-axis [by the saddle nature of $(K,0)$] in which the flow is away from $(K,0)$ (see Fig. 4.2).

We now come to the all important equilibrium (iii). Setting $x = x^*$ and $y = y^*$ and using (4.12) and (4.13), we get

$$J(x^*,y^*) = \begin{bmatrix} H(x^*) & -p(x^*) \\ y^*q_x(x^*) & 0 \end{bmatrix} \tag{4.19}$$

where

$$H(x^*) = x^*g_x(x^*) + g(x^*) - \frac{x^*g(x^*)p_x(x^*)}{p(x^*)} \tag{4.20}$$

Since the eigenvalues of $J(x^*,y^*)$ are given by

$$\frac{H(x^*)}{2} \pm \frac{\left[H(x^*)^2 - 4y^*p(x^*)q_x(x^*)\right]^{1/2}}{2} \tag{4.21}$$

it is clear that the sign of the real parts of these eigenvalues coincide with the sign of $H(x^*)$. From this we may come to the following conclusions, again utilizing (4.13).

$$H(x^*)^2 - 4x^*g(x^*)q_x(x^*) \begin{matrix} < 0 \\ > 0 \end{matrix} \Rightarrow (x^*,y^*) \begin{matrix} \text{spiral point} \\ \text{or center node} \end{matrix} \tag{4.22}$$

Further,

$$H(x^*) \begin{matrix} < 0 \\ > 0 \end{matrix} \Rightarrow (x^*,y^*) \begin{matrix} \text{stable} \\ \text{unstable} \end{matrix} \tag{4.23}$$

The case where $H(x^*) = 0$ is undecided, in which case (x^*,y^*) could be either a center or a stable spiral point or an unstable spiral point. We note, of course, that for the Lotka-Volterra system $H(x^*) = 0$ and (x^*,y^*) is a center.

Several graphical techniques have been developed to determine the stability of (x^*,y^*). Such techniques may lend themselves to analysis of the stability based on experimental data. These techniques will be discussed in the next two sections.

4.4 THE METHOD OF ROSENZWEIG AND MACARTHUR

Before discussing (and verifying) the first of the graphical methods for determining the stability of (x^*,y^*), we need to discuss the isoclines of system (4.2), which we now define.

DEFINITION 4.1 The curve in the interior of the first quadrant of the x-y plane along which the prey growth is instantaneously zero (i.e., $dx/dt = 0$) is called the *prey isocline*. The curve along which the predator growth is instantaneously zero is called the *predator isocline*.

The predator isocline is trivial to graph. It is simply the vertical line $x = x^*$. The prey isocline is more interesting, being given by the graph of

$$y = \frac{xg(x)}{p(x)} \tag{4.24}$$

Clearly, this isocline goes through the three points $(0,\alpha/\beta)$, (x^*,y^*), and $(K,0)$ (see Fig. 4.3). Of course, the intersection of the two isoclines is always at equilibrium (iii). If $\alpha/\beta \geq y^*$, this isocline may or may not have local maxima and/or minima. If $\alpha/\beta < y^*$, the prey isocline must have at least one local maximum. In Rosenzweig (1969) empirical ecological arguments are given to explain why the prey isocline should have at least one local maximum.

Several of the possibilities described above are shown in Fig. 4.3.

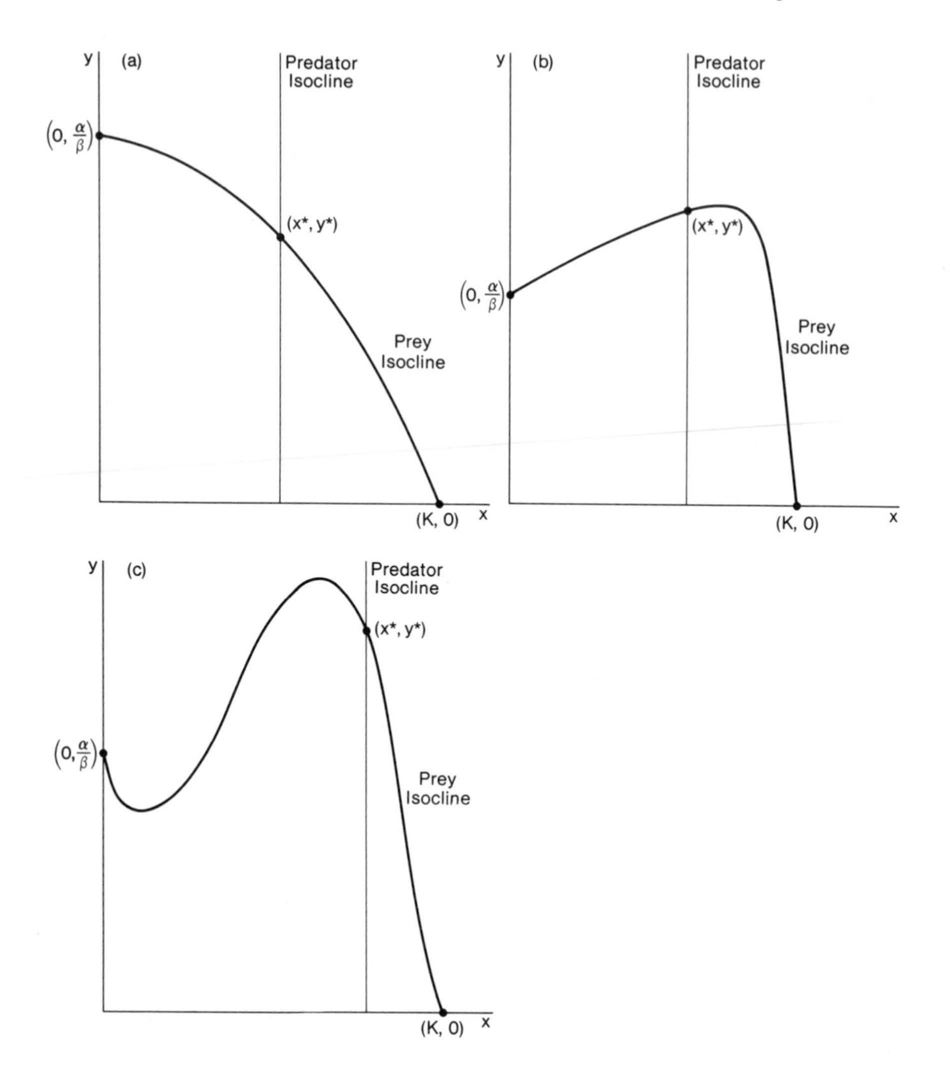

Fig. 4.3 (a) $\alpha/\beta > y^*$. No local maxima occur. (b) $\alpha/\beta < y^*$. One local maximum occurs, but there are no local minima. (c) $\alpha/\beta < y^*$. There is both a local maximum and a local minimum.

We now proceed to derive the results of Rosenzweig and MacArthur (1963). We consider again $H(x^*)$ as given by (4.20) and note that since both x^* and $g(x^*)$ are positive, $H(x^*)/x^*g(x^*)$ has the same sign as $H(x^*)$. From (4.20),

$$\frac{H(x^*)}{x^*g(x^*)} = \frac{g_x(x^*)}{g(x^*)} + \frac{1}{x^*} - \frac{p_x(x^*)}{p(x^*)} = \frac{d}{dx} \ell n \left[\frac{xg(x)}{p(x)}\right]\Bigg|_{x=x^*} \tag{4.25}$$

Then by (4.25), criterion (4.23) becomes

$$\frac{d}{dx} \ell n \left[\frac{xg(x)}{p(x)}\right]\Bigg|_{x=x^*} \begin{matrix} < 0 \\ > 0 \end{matrix} \Rightarrow (x^*,y^*) \begin{matrix} \text{stable} \\ \text{unstable} \end{matrix} \tag{4.26}$$

However, since the derivative of a function is positive or negative if and only if the function itself is increasing or decreasing respectively, and since the logarithm is an increasing function, (4.26) is equivalent to

$$\frac{xg(x)}{p(x)} \begin{matrix} \text{decreasing} \\ \text{increasing} \end{matrix} \text{ at } x^* \Rightarrow (x^*,y^*) \begin{matrix} \text{stable} \\ \text{unstable} \end{matrix} \tag{4.27}$$

This is the criterion of Rosenzweig and MacArthur (1963).

Looking at this criterion graphically, (4.27) says that if the prey and predator isoclines intersect at a point of decrease of the prey isocline (perhaps just to the right of a local maximum), the equilibrium is asymptotically stable. However, if the intersection is at a point of increase (perhaps just to the left of a local maximum), the equilibrium is unstable. In Fig. 4.3 we see that the equilibria in (a) and (c) are stable, whereas the equilibrium in (b) is unstable.

In Rosenzweig and MacArthur (1963) the case where the predator isocline was not just a vertical line was considered. The reader should be cautioned that in that case the sign of the real parts of the eigenvalues of $J(x^*,y^*)$ is not necessarily given by $H(x^*)$ and the preceding analysis is not necessarily valid (i.e., Rosenzweig and MacArthur's criterion may not hold).

A second graphical criterion is developed in the next section.

4.5 THE METHOD OF GAUSE, SMARAGDOVA, AND WITT

In this case we consider

$$\frac{p(x^*)H(x^*)}{g(x^*)} = p(x^*) - x^*p_x(x^*) + \frac{x^*p(x^*)g_x(x^*)}{g(x^*)} \tag{4.28}$$

Then since $p(x^*)H(x^*)/g(x^*)$ has the same sign as $H(x^*)$, criterion (4.23) is equivalent to

$$p(x^*) - x^*p_x(x^*) \lessgtr \frac{-x^*p(x^*)g_x(x^*)}{g(x^*)} \Rightarrow (x^*,y^*) \begin{array}{l}\text{stable}\\ \text{unstable}\end{array} \tag{4.29}$$

For convenience of notation we let y_0 denote the right side of the inequality in (4.29), i.e.,

$$y_0 = \frac{-x^*p(x^*)g_x(x^*)}{g(x^*)} \tag{4.30}$$

We now give a geometrical interpretation to $p(x^*) - x^*p_x(x^*)$. The slope of the tangent line to the predation curve $y = p(x)$ at x^* is just $p_x(x^*)$, and hence the equation of this line is

$$y = (x - x^*)p_x(x^*) + p(x^*) \tag{4.31}$$

Let $\bar{y}$ be the y-intercept of this tangent line; then from (4.31)

$$\bar{y} = p(x^*) - x^*p_x(x^*) \tag{4.32}$$

From this is derived the criterion first hinted at in Gause et al. (1936):

$$\bar{y} \lessgtr y_0 \Rightarrow (x^*,y^*) \begin{array}{l}\text{stable}\\ \text{unstable}\end{array} \tag{4.33}$$

This criterion tends to lend some credence to statements that certain types of predator functional responses are stabilizing or destabilizing. Some of these aspects are illustrated in Fig. 4.4. In the first instance if $y_0 = 0$ [i.e., $g_x(x^*) = 0$], then a Holling type of predator response is always unstable. This is in fact what

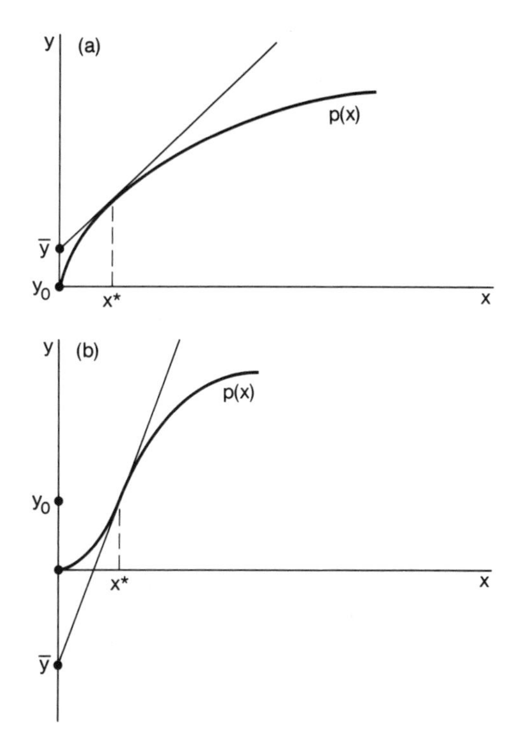

Fig. 4.4 (a) $\bar{y} > y_0 = 0$. (x^*,y^*) unstable. (b) $\bar{y} < y_0$. (x^*,y^*) stable.

was noticed by Gause, Smaragdova, and Witt and rediscovered by Oaten and Murdoch (1975) (see Fig. 4.4a). If there is a sigmoid type of functional response, so that x^* is such that $\bar{y} < 0$, then the equilibrium is stable (since $y_0 \geq 0$) (see Fig. 4.4b).

The method then consists of plotting y_0 on a y-axis, graphing the predation curve $y = p(x)$, and drawing the tangent line to this curve at x^*. If the y-intercept $\bar{y}$ lies above y_0, then (x^*,y^*) is unstable. If it lies below y_0, then the equilibrium is stable. If $\bar{y} = y_0$, then there could occur stability or instability or neutral stability.

4.6 PHASE PLANE ANALYSIS

At this time we give a general phase plane analysis for system (4.2). This analysis is carried out under the assumption that a carrying capacity exists. We illustrate this analysis in Fig. 4.5. We divide the first quadrant of the phase plane into five zones and their boundaries as described below:

$$\text{Zone I} = \left\{(x,y):\ x^* < x < K,\quad y > \frac{xg(x)}{p(x)}\right\}$$

$$\text{Zone II} = \left\{(x,y):\ 0 < x < x^*,\quad y > \frac{xg(x)}{p(x)}\right\}$$

$$\text{Zone III} = \left\{(x,y):\ 0 < x < x^*,\quad 0 < y < \frac{xg(x)}{p(x)}\right\}$$

$$\text{Zone IV} = \left\{(x,y):\ x^* < x < K,\quad 0 < y < \frac{xg(x)}{p(x)}\right\}$$

$$\text{Zone V} = \{(x,y):\ K < x,\quad 0 < y\}$$

We analyze the phase plane by considering dy/dx, which is obtained from system (4.2) be eliminating t. This gives

$$\frac{dy}{dx} = \frac{y[-\gamma + q(x)]}{xg(x) - yp(x)} \tag{4.34}$$

We first consider dy/dx on the isoclines. On the predator isocline, dy/dx = 0, and so solutions crossing the predator isocline do so horizontally. Further, since $dx/dt > 0$ for $y < xg(x)/p(x)$, and

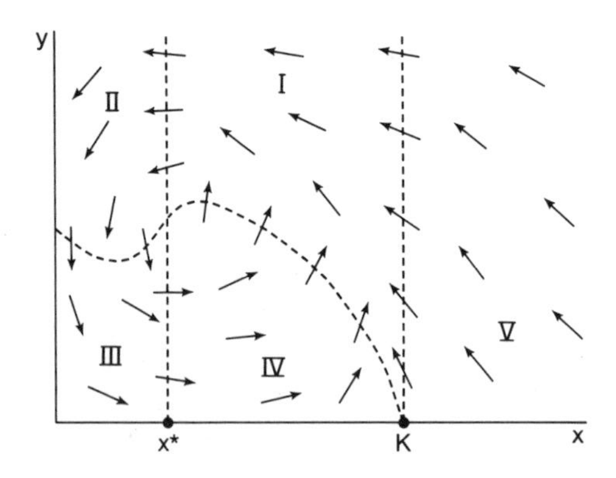

Fig. 4.5 Flow direction in the phase plane for system (4.2).

$dx/dt < 0$ for $y > xg(x)/p(x)$, solutions cross from zone I to zone II and from zone III to zone IV. On the prey isocline $dx/dy = 0$, and so solutions cross the prey isocline vertically. Also, since $dy/dt < 0$ for $x < x^*$ and $dy/dt > 0$ for $x > x^*$, solutions cross from zone II to zone III, and from zone IV to zone I.

We now consider dy/dx on the line $x = K$, which, since $g(K) = 0$, reduces to

$$\left.\frac{dy}{dx}\right|_{x=K} = \frac{-\gamma + q(K)}{-p(K)} < 0 \tag{4.35}$$

and is independent of y. Also, since $dx/dt|_{x=K} < 0$, all solutions crossing $x = K$ do so at a constant negative slope from zone V to zone I.

It is easy to see as well that $dy/dx < 0$ in zones I and III, whereas $dy/dx > 0$ in zones II and IV. Also $dy/dx < 0$ in zone V. It is also clear that $dx/dt < 0$ in zones I, II, and V, whereas $dx/dt > 0$ in zones III and IV. This completes the flow patterns as illustrated in Fig. 4.5.

Now consider a solution initiating in zone V. Writing (4.34) as

$$\frac{dy}{dx} = \frac{-\gamma + q(x)}{xg(x)/y - p(x)} \tag{4.36}$$

then for fixed $x > K$, we have

$$-\frac{-\gamma + q(x)}{p(x)} < \frac{dy}{dx} < 0 \tag{4.37}$$

and since solutions move to the left,

$$-\sup_{K \leq u \leq x} \frac{-\gamma + q(u)}{p(u)} \leq \frac{dy}{dx} \leq 0 \tag{4.38}$$

so long as the trajectory remains to the right of K. This implies that all trajectories initiating in zone V flow to the left with bounded slopes and hence must cross $x = K$ into zone I in finite time.

Consider now such a solution crossing $x = K$ into zone I at $y = y_1$, where $y_1 > \max_{0 \leq x \leq K} xg(x)/p(x)$. Let this trajectory be labeled Γ. Then Γ continues to the left with a negative slope. From (4.3)

$$-\sup_{\substack{x^* \leq x \leq K \\ y \geq y_1}} \left| \frac{-\gamma + q(x)}{xg(x)/y - p(x)} \right| \leq \frac{dy}{dx} \leq 0 \tag{4.39}$$

for Γ so long as it remains in zone I. Hence Γ must cross the predator isocline in finite time into zone II. Here Γ continues to the left but with a positive slope. Since it cannot intersect the y-axis, it must cross the prey isocline into zone III. From there it must cross into zone IV, since again it cannot intersect the x-axis. Now, since it once more has positive tangential slope and the equilibrium at (K,0) is hyperbolic, the trajectory crosses into zone I below y_1; hence, since Γ cannot intersect itself, the inward spiraling must continue. This is illustrated in Fig. 4.6.

It now follows from Chap. 16 of Coddinton and Levinson (1955) that the positive limit set of Γ, $L(\Gamma^+)$, is a nonempty, closed, and connected set. This limit set could be the equilibrium (x^*,y^*).

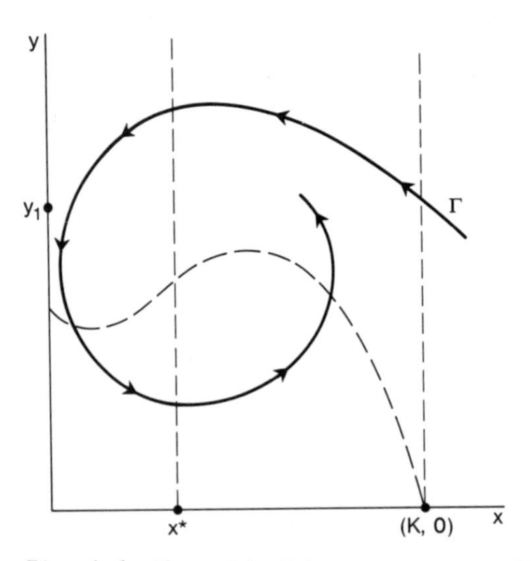

Fig. 4.6 The orbit Γ has a compact ω limit set.

If not, it must be a periodic orbit, which is at least stable from the outside. In any case, there is always an invariant attracting set. Also if (x^*,y^*) is unstable, then $L(\Gamma^+)$ must be a limit cycle. In any case, system (4.2) always has either a stable equilibrium or a stable from the outside limit cycle, or both.

4.7 SYSTEMS WITH CONSTANT SOURCE TERMS

In this section system (4.2) is modified to include a source term in the prey equation. This modified system is of the form

$$\begin{aligned} x' &= xg(x) - yp(x) + r \\ y' &= y[-\gamma + q(x)] \end{aligned} \tag{4.40}$$

Here r may be positive or negative. A negative source may be thought of as representing a harvesting effect.

In the absence of predators the prey species grows according to the growth law

$$x' = xg(x) + r \tag{4.41}$$

(see Chap. 1, Exercises). For $r < 0$, small initial populations could lead to extinction in finite time. In any case, there continues to be a flow on the x-axis. However, the same is not true on the y-axis, since $x = 0$ is instantaneous only (i.e., all solutions touching the y-axis cross into the first quadrant in the case $r > 0$, and out of the first quadrant in the case $r < 0$).

Consider now the question of an interior equilibrium. The x-coordinate of this equilibrium, x^*, remains unchanged. Let $\hat{y}$ be the new coordinate of the equilibrium. Then

$$\hat{y} = \frac{x^*g(x^*) + r}{p(x^*)} = y^* + \frac{r}{p(x^*)} \tag{4.42}$$

Hence if $r > 0$, the equilibrium is vertically higher in the plane, whereas if $r < 0$, it is lower. If $r < 0$ and $|r| > x^*g(x^*)$, then a positive equilibrium no longer exists.

The question occurs: How does the presence of sources affect the stability of the equilibrium?

Reexamining the considerations that led to y_0 as defined in (4.30) and $\bar{y}$ as defined in (4.32), we see that the new criteria for stability and instability are that

$$\bar{y} \lessgtr \hat{y}_0 \Rightarrow (x^*,\hat{y}) \begin{array}{l}\text{stable}\\ \text{unstable}\end{array} \tag{4.43}$$

where

$$\begin{aligned}\hat{y}_0 &= \frac{-x^*p(x^*)g_x(x^*)}{g(x^*)} + \frac{rp_x(x^*)}{g(x^*)} \\ &= y_0 + \frac{rp_x(x^*)}{g(x^*)}\end{aligned} \tag{4.44}$$

Hence $\hat{y}_0 > y_0$ if $r > 0$ and $\hat{y}_0 < y_0$ if $r < 0$. Since the effect of raising the y_0 to $\hat{y}_0$ is to stabilize and the effect of lowering is to destabilize, we get the results that spraying the system with prey is a stabilizing process, whereas harvesting the prey is a destabilizing process. This is of particular importance in applications to pest control in Chap. 6.

The preceding considerations are illustrated in Fig. 4.7 and 4.8.

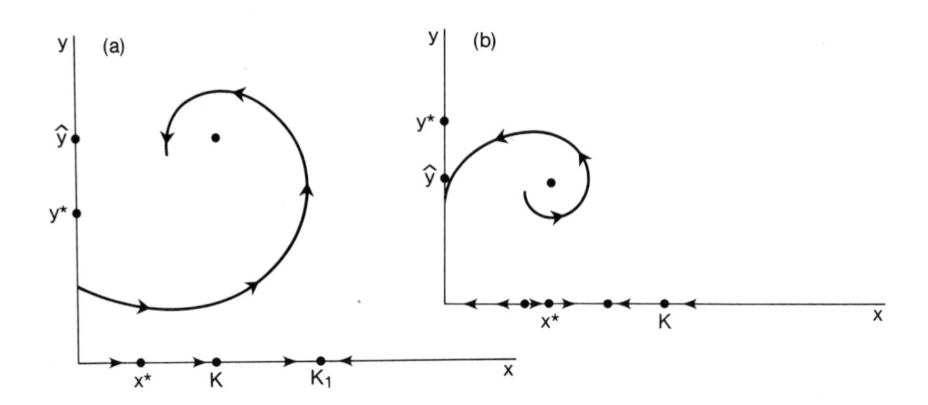

Fig. 4.7 (a) An orbit resulting from a system with a positive source in the prey equation. (b) An orbit resulting from a system with a negative source in the prey equation.

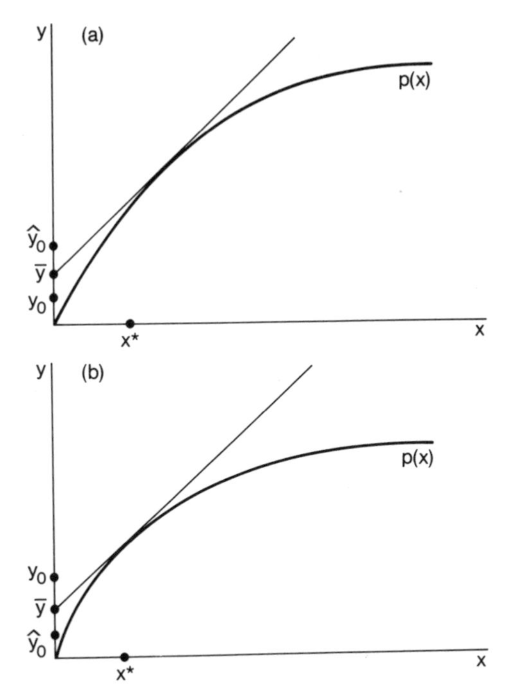

Fig. 4.8 (a) The effect of r on stability: $r > 0$. (b) The effect of r on stability: $r < 0$.

NOTES ON THE LITERATURE

Oscillations with and without extinctions have been observed in many laboratory and natural situations. For some references see Carbyn (1974), Force (1974), Gause et al. (1936), Jenson and Ball (1970), Luckinbill (1974), McAllister et al. (1972), Sudo et al. (1975), van den Ende (1973).

The Rosenzweig and MacArthur isocline method has been utilized by ecologists and others to predict the stability of predator-prey systems. See, for example, Maly (1975), Noy-Meir (1975). Strebel and Goel (1973) and Tanner (1975) have given theoretical discussions on the Rosenzweig and MacArthur method. Rosenzweig (1973a) has utilized his method to discuss evolutionary behavior of a predator-prey system. Oaten and Murdoch (1975a,b) also discuss the graphical method of Gause, Smaragdova, and Witt, as does Freedman (1976). See also Armstrong (1976) for yet another graphical method for stability.

De Angelis (1975c) considers estimating the predator-prey limit cycle. Foster and Rapoport (1956) treat a predator-prey system from a game-theoretic point of view. Hubbell (1973b) looks at a predator-prey system with oscillations due to variations in energy flows. Levin (1972) has considered genetic change in the prey due to predator pressure. Montroll (1972) considers some statistical aspects. Schoener (1973) considers population regulation due to energy or time. Utz and Waltman (1963) consider the existence of a periodic solution for a certain system of intermediate type.

EXERCISES

4A.1 Analyze system (4.2) where y in the prey equation is replaced by y^{1-m}, where the mutual interference constant m is such that $0 < m < 1$. In particular, how does m affect the interior equilibrium and its stability?

4A.2 Analyze the phase plane flows for system (4.2) in the case where there is no carrying capacity.

4A.3 (Open problem) Analyze system (4.39) in the case where r is a periodic function of t. There will no longer be an interior equilibrium. What can one say about the phase plane? For instance, will there continue to be an interior periodic orbit?

4A.4 Consider a modification of system (4.2) where a source term is added to the predator equation. How does this affect the analysis?

B: DISCRETE MODELS

There are certain predator-prey systems that are more naturally described by difference than by differential equations. Such models are termed discrete models. In the remainder of this chapter we describe and analyze such discrete systems, including the discrete analogue of the model with a Holling-type predation and mutual interference.

4.10 INTRODUCTION

As mentioned in Section 1.6, there are times when it is more natural to consider discrete models than to consider continuous models. If the population numbers are small or if births and deaths all occur at discrete times, or within certain intervals of time, such as a generation, discrete models would indeed be more realistic. At this time we shall develop a discrete predator-prey model. Here t will denote discrete times (perhaps years or generations).

The assumptions made on the prey are that the number of prey at time t + 1 is equal to the number of prey at time t, plus the number of prey born from t to t + 1, minus the number of deaths due to natural causes, minus the number of prey attacked (and killed) by predators. The assumptions made about the predators are that the number of predators at time t + 1 is equal to the number of predators at time t, plus the number born, minus the number that die. In addition, it is assumed that the birth rate and natural death rate of the prey and the death rate of the predators are constants.

Using these assumptions, the model is

$$\begin{aligned} x(t+1) &= (1+r)x(t) - x_a(t) \\ y(t+1) &= (1-d)y(t) + y_b(t) \end{aligned} \qquad (4.45)$$

where r is the birth rate minus natural death rate of the prey, d is the death rate of the predators, $x_a(t)$ is the number of prey devoured, and $y_b(t)$ is the number of predators born. In the case of certain predators, such as some parasitoids, $d = 1$, since all members of the generation may die at the end of the season. In general, however, $r \geq 0$, $0 < d \leq 1$.

We now make our next assumption, similar to one used by Gause et al. (1936) in setting up their continuous model, namely, that the number of predators born is proportional to the number of prey devoured. Putting this in system (4.45) leads to the model

$$\begin{aligned} x(t + 1) &= (1 + r)x(t) - x_a(t) \\ y(t + 1) &= (1 - d)y(t) + cx_a(t) \end{aligned} \tag{4.46}$$

where $0 < c < 1$.

At this time, we shall discuss several candidates for $x_a(t)$. The first was considered as applicable by Dixon and Cornwell (1970) and the model was analyzed by them in their paper utilizing computer simulations based on data obtained from moose and wolf interactions on Isle Royale. The formula for $x_a(t)$ itself was developed by Watt (1959) and is given by

$$x_a(t) = Ky(t)\{1 - \exp[-kx(t)^2 y(t)^{1-m}]\} \tag{4.47}$$

This expression, depending on k, is close to Ky(t) and could be replaced by Ky(t) to estimate the behavior of the model (4.45); m is the mutual interference constant, as described in Table 2.1, and $0 \leq m < 1$.

A second expression for $x_a(t)$ was developed by Hassell and Rogers (1972) and has the form

$$x_a(t) = \frac{\alpha x(t)y(t)^{1-m}}{1 + \beta x(t)} \tag{4.48}$$

where the positive constants α and β are related to the handling time, attack coefficient, and so on. Again $0 \leq m < 1$. We will utilize this last expression for $x_a(t)$ in our discrete model. Substituting (4.48) into (4.46), we get

$$\begin{aligned} x(t + 1) &= (1 + r)x(t) - \frac{\alpha x(t)y(t)^{1-m}}{1 + \beta x(t)} \\ y(t + 1) &= (1 - d)y(t) + \frac{c\alpha x(t)y(t)^{1-m}}{1 + \beta x(t)} \end{aligned} \tag{4.49}$$

This system will be analyzed in the following sections.

4.11 THE CASE m = 0

We consider now the special case $m = 0$, the simplest case to handle. System (4.49) becomes

$$\begin{aligned} x(t+1) &= (1+r)x(t) - \frac{\alpha x(t)y(t)}{1+\beta x(t)} \\ y(t+1) &= (1-d)y(t) + \frac{c\alpha x(t)y(t)}{1+\beta x(t)} \end{aligned} \tag{4.50}$$

We now proceed to show that under reasonable biological assumptions, a steady state for this system exists. Of course by a steady state we mean an $x^*(t)$ and $y^*(t)$ such that

$$x^*(t+1) = x^*(t), \qquad y^*(t+1) = y^*(t) \tag{4.51}$$

To find the steady state, substitute (4.51) into (4.50); a straightforward calculation gives

$$x^*(t) = \frac{d}{c\alpha - d\beta}, \qquad y^*(t) = \frac{cr}{c\alpha - d\beta} \tag{4.52}$$

For this steady state to be meaningful, both $x^*(t)$ and $y^*(t)$ must be positive. This is clearly the case, provided that

$$c\alpha - d\beta > 0 \tag{4.53}$$

If (4.53) were reversed, then writing the second part of Eq. (4.50) as

$$\begin{aligned} y(t+1) &= \frac{y(t)}{1+\beta x(t)}\,[1 - d + (\beta + c\alpha - d\beta)x(t)] \\ &= y(t)\left[1 + \frac{(c\alpha - d\beta)x(t) - d}{1+\beta x(t)}\right] \end{aligned} \tag{4.54}$$

and $y(t+1) < y(t)$. Biologically, then, since $[(c\alpha - d\beta)x(t) - d]/[1 + x(t)] \leq -\ell^2 < 0$, for $x(t) \geq 0$, extinction must occur (perhaps even in finite time). Hence (4.53) must be valid biologically for the system to persist.

Now, to examine the stability of the steady state, write system (4.50) as

$$x(t+1) - x(t) = rx(t) - \frac{\alpha x(t)y(t)}{1+\beta x(t)}$$

$$y(t+1) - y(t) = -dy(t) + \frac{c\alpha x(t)y(t)}{1 + \beta x(t)} \tag{4.55}$$

Then the variational matrix of the right side is

$$J(x,y) = \begin{bmatrix} r - \dfrac{\alpha y(t)}{[1 + \beta x(t)]^2} & -\dfrac{\alpha x(t)}{1 + \beta x(t)} \\ \dfrac{c\alpha y(t)}{[1 + \beta x(t)]^2} & -d + \dfrac{c\alpha x(t)}{1 + \beta x(t)} \end{bmatrix} \tag{4.56}$$

Evaluating at the steady state gives

$$J[x^*(t),y^*(t)] = \begin{bmatrix} \dfrac{dr\,\beta}{c\alpha} & -\dfrac{d}{c} \\ \dfrac{r(c\alpha - d\beta)}{\alpha} & 0 \end{bmatrix} \tag{4.57}$$

Then the eigenvalues of this matrix are

$$\lambda = \frac{dr\beta}{2c\alpha} \pm \frac{1}{2}\left[\frac{d^2r^2\beta^2}{c^2\alpha^2} - \frac{4dr(c\alpha - d\beta)}{c\alpha}\right]^{1/2} \tag{4.58}$$

Hence the sign of the real part of λ is given by the sign of $dr\beta/2c\alpha > 0$, thus the steady state is unstable.

To examine this further, let

$$x(t) = x^* + X, \qquad y(t) = y^* + Y \tag{4.59}$$

Then substituting from (4.59) and (4.52) into (4.50) and simplifying gives

$$x(t+1) = x(t)\left[1 + \frac{r\beta X - \alpha Y}{1 + \beta x(t)}\right] \tag{4.60}$$

and

$$y(t+1) = y(t)\left[1 + \frac{(c\alpha - d\beta)X}{1 + \beta x(t)}\right] \tag{4.61}$$

From (4.61) it is clear that if $x(t) > x^*$, then $y(t+1) > y(t)$, and if $x(t) < x^*$, then $y(t+1) < y(t)$. Hence, even if $Y = 0$, $y(t+1) \neq y(t)$ if $X \neq 0$, no matter how small. In this sense the steady state is unstable.

There could, however, occur oscillations. That is clear from (4.60) and (4.61). If $X > 0$ and $Y > 0$, then if αY is greater than $r\beta X$ by a substantial amount, it could be that, even though $y(t + 1) > y(t) > y^*$, $x(t + 1) < x^* < x(t)$. Then $x(t + 2) < x(t + 1)$ and $y(t + 2) < y(t + 1)$. If Y is eventually dragged below zero sufficiently, then it could cause X to increase, after which it could decrease again, then increase again, etc.

In the next section we consider the case $m > 0$.

4.12 THE CASE 0 < m < 1

We consider again system (4.49) when $0 < m < 1$. We again seek a steady state. This is equivalent to finding x^* and y^*, so that

$$r - \frac{\alpha y^{*1-m}}{1 + \beta x^*} = -d + \frac{c\alpha x^* y^{*-m}}{1 + \beta x^*} = 0 \tag{4.62}$$

or on substituting $1 + \beta x^* = (\alpha/r)y^{*1-m}$ into the last equality of (4.62), we get, on simplifying,

$$dy^* = rcx^* \tag{4.63}$$

Substituting from (4.63) into $r - (\alpha y^{*1-m})/(1 + \beta x^*) = 0$ and again simplifying, we get the equation

$$f(y^*) = 0 \tag{4.64}$$

where

$$f(y^*) = d\beta y^* - c\alpha y^{*1-m} + cr \tag{4.65}$$

Since $f(0) = cr > 0$ and $\lim_{y^*\to\infty} f(y^*) = +\infty$, there will be a positive steady state if and only if the minimum over $y^* > 0$ of $f(y^*)$ is nonpositive. This minimum is obtained, of course, by setting $df(y^*)/dy^* = 0$ and solving for y^*_{min}, which gives

$$y^*_{min} = \left[\frac{c\alpha(1 - m)}{d\beta}\right]^{1/m} \tag{4.66}$$

Then on substituting (4.66) into (4.65), the condition for a positive steady state becomes

$$cr - \frac{d\beta m}{1 - m}\left[\frac{c\alpha(1 - m)}{d\beta}\right]^{1/m} \leq 0 \tag{4.67}$$

After rearranging, raising to the m-th power, and simplifying, condition (4.67) reduces to

$$d^{1-m}r^m\beta^{1-m} \leq c^{1-m}m^m(1 - m)^{1-m}\alpha \tag{4.68}$$

We note that in the case of strict inequality, if m = 0, (4.68) reduces to (4.53).

Assume that (4.68) holds for the remainder of this section. If equality holds, there will occur a single positive steady state given by $y = y^*_{min}$, $x = dy^*_{min}/cr$. If inequality holds, there will be two distinct steady states, given by $y = y^*_1$, $x = dy^*_1/cr$ and $y = y^*_2$, $x = dy^*_2/cr$, where $0 < y^*_1 < y^*_{min} < y^*_2$. The analysis of this model beyond the existence of a steady state is of some difficulty, and to the best of this author's knowledge has not been done.

The graphs of typical curves $y^{*1-m} = (r/\alpha)(1 + \beta x^*)$ and $y^{*m} = c\alpha/d[x^*/(1 + \beta x^*)]$ are shown in Fig. 4.9. They indicate how intersection giving two steady states can occur.

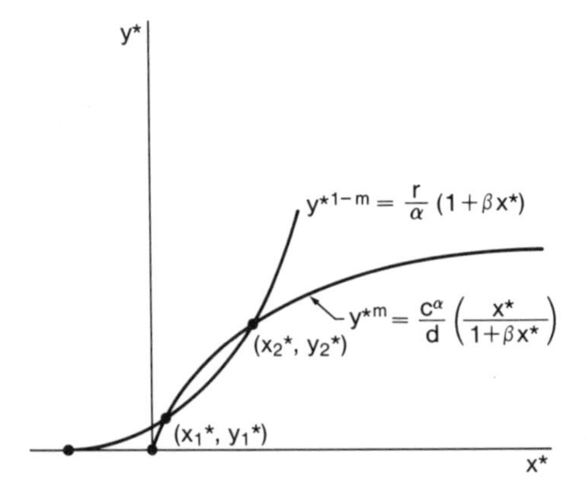

Fig. 4.9 Isoclines of the discrete system (4.49) with m = 1/2, r = α, d = cα, β small.

NOTES ON THE LITERATURE

For other analyses and derivations involving discrete deterministic models, see J. Allen (1975), Bodenheimer and Schiffer (1952), Levine (1975), Maynard Smith and Slatkin (1973), Murdie and Hassell (1973). For stochastic discrete models, see Diamond (1974b), Leslie (1958), Leslie and Gower (1960), Poole (1974).

EXERCISES

4B.1 (Open but solvable problem) Determine the stability of the steady states deduced in Sec. 4.11. Determine possible behaviors of the solutions.

4B.2 Analyze system (4.46) with $x_a(t)$ given by (4.47) (see Dixon and Cornwell, 1970).

4B.3 (Open) Develop and analyze a discrete model incorporating a carrying capacity in the environment (i.e., a model with a nonconstant prey birth rate and natural death rate).

Chapter 5
THE KOLMOGOROV MODEL

5.1 INTRODUCTION

The Kolmogorov model of growth is, mathematically, the most general model of the type considered in this book incorporating the principle that the growth rate of a species is proportional to the number of the species present. Hence, in the first instance, the model is given by

$$\begin{aligned} x' &= xf(x,y) \\ y' &= yg(x,y) \end{aligned} \tag{5.1}$$

Of course, conditions must be put on f and g to make x a prey and y a predator. This will be done in the next two sections, at which time an analysis of the model will be given. Following that, perturbed models will be considered, for which it will not always be assumed that the two species necessarily have a predator-prey relationship.

5.2 KOLMOGOROV'S MODEL

The model first developed by Kolmogorov (1936) in 1936 -- and expanded on by Rescigno and Richardson (1965), May (1972), and Albrecht et al. (1973) -- is given by the system of differential equations (5.1) subject to the following hypotheses. The hypotheses, which are assumed to hold for $x \geq 0$, $y \geq 0$, and the reasoning behind them are as follows.

1. First it is assumed that for a fixed number of prey, the prey growth rate is diminished by increasing the number of predators. This leads to the condition

$$\frac{\partial f(x,y)}{\partial y} < 0 \tag{5.2}$$

2. For a fixed ratio of predators to prey it is assumed that increasing the number of predators tends to diminish the growth rate of the prey, since then the chance of encounter between predator and prey are increased. This says that the change in f along the outward vector from the origin is negative, or

$$x \frac{\partial f(x,y)}{\partial x} + y \frac{\partial f(x,y)}{\partial y} < 0 \tag{5.3}$$

3. For small populations of both predator and prey it is assumed that the prey population increases, and hence

$$f(0,0) > 0 \tag{5.4}$$

4. There must be some sufficiently large value of the number of predators at which a small prey population can no longer increase. Hence, to preserve continuity,

$$\exists A > 0 \ni f(0,A) = 0 \tag{5.5}$$

5. The next condition corresponds to the existence of a carrying capacity of the environment, i.e.,

$$\exists B > 0 \ni f(B,0) = 0 \tag{5.6}$$

6. Since the predators are competing for the same resource, holding the number of prey fixed and increasing the number of predators will tend to slow the predator growth rate. This gives

$$\frac{\partial g(x,y)}{\partial y} < 0 \tag{5.7}$$

7. However, for the same reasons that lead to (5.3), we get

$$x \frac{\partial g(x,y)}{\partial x} + y \frac{\partial g(x,y)}{\partial y} > 0 \tag{5.8}$$

8. If there are sufficiently many prey, the predator growth rate will be positive, but if there are too few prey, even small predator populations will have a negative growth rate. Hence

$$\exists C > 0 \ni g(C,0) = 0 \tag{5.9}$$

9. Finally, as stated by both Kolmogorov (1936) and Rescigno and Richardson (1965), if $B \leq C$, the predators will always go extinct (and note that that is true for the intermediate models). Hence

$$B > C \tag{5.10}$$

The conclusions reached by Kolmogorov (1936) and Rescigno and Richardson (1965) were that there is either a stable interior equilibrium, or a stable limit cycle, or both. These conclusions were not proved. The arguments used were graphical, and it is clear from the graphs (see Fig. 5.1) that additional assumptions were made, namely, that the isoclines [i.e., the curves $f(x,y) = g(x,y) = 0$] are essentially as in Fig. 5.1. However, these assumptions were not proved utilizing hypotheses 1 to 9. It will be shown in the next section that with slight modifications, on the given assumptions, the conclusions are nearly correct.

A second fault with the model as it now stands is an inconsistency in (5.7) and (5.8). If we let $x = 0$, $y > 0$, then (5.7) and (5.8) become $\partial g(0,y)/\partial y < 0$, $y[\partial g(0,y)/\partial y] > 0$, which is self-contradictory. This was first noticed by May (1972), who suggested modifying the model so as to allow equality to zero on the y-axis. In such a case, however, it was pointed out by Albrecht et al. (1973) that there may be a region of neutral stability (i.e., an annular region filled with periodic solutions).

5.3 PROOF OF A KOLMOGOROV-TYPE THEOREM

The material of this section is due to Albrecht et al. (1974).

First, in addition to hypotheses 1 to 9 on system (5.1), it is assumed that (5.2), (5.3), (5.7), and (5.8) need hold in the

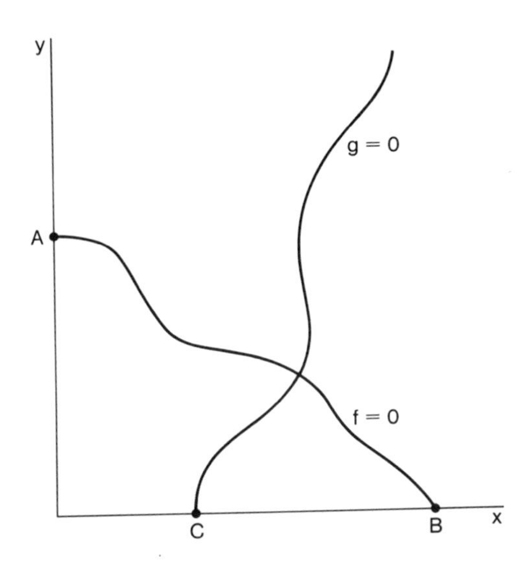

Fig. 5.1 A graph of the isoclines of model (5.1) similar to those appearing in Kolmogorov (1936) and Rescigno and Richardson (1965).

interior of the first quadrant only, that the sign $<$ in (5.7) may be replaced by $\leq$, and that

$$
\begin{aligned}
&10.\quad (x - B)f(x,0) < 0\\
&11.\quad (y - A)f(0,y) < 0\\
&12.\quad (x - C)g(x,0) > 0
\end{aligned}
\tag{5.11}
$$

This guarantees that the equilibria described on the axes are the only such equilibria. We now state and prove a Kolmogorov-type theorem.

THEOREM 5.1 *Let hypotheses* 1 *to* 12 *hold. Then system* (5.1) *has a unique equilibrium interior to the first quadrant. If this equilibrium is not asymptotically stable, there is a limit cycle interior to the first quadrant which is asymptotically stable from the outside.*

PROOF. During the course of this proof the reader is referred to Fig. 5.2. First we show the existence of a continuous function

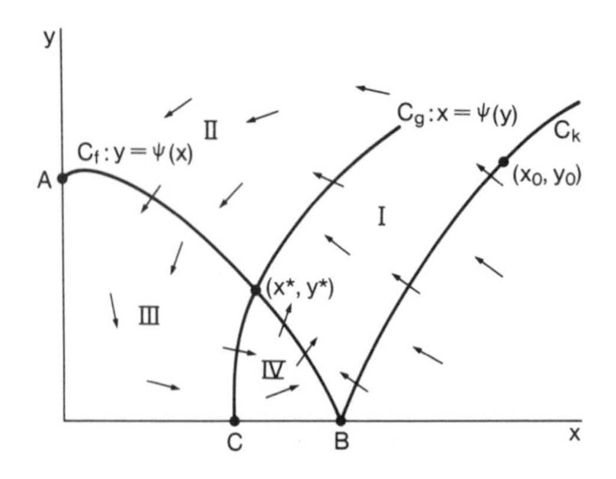

Fig. 5.2 Direction of the orbits under the hypotheses of Theorem 5.1.

$y = \phi(x)$, such that

$$f[x,\phi(x)] = 0, \qquad \phi(0) = A, \qquad \phi(B) = 0 \tag{5.12}$$

$\phi(x)$ is obtained by solving $f(x,y) = 0$ for y as a function of x. Clearly, for each fixed $x \geq 0$, there is at most one $y > 0$ such that $f(x,y) = 0$. For $x = 0$ this follows from 11, and for $x > 0$, from (5.2). If $x > B$, then by 10, $f(x,0) < 0$ and hence the domain of $\phi(x)$ is contained in $[0,B]$. Now from (5.5) $\phi(0) = A$. By continuity, there is a $\delta > 0$ such that for $0 \leq x \leq \varepsilon$, $f(x, A + \delta) < 0$ and $f(x, A - \delta) > 0$. Hence, in particular, in this interval [i.e., $0 \leq x \leq \varepsilon$] $\phi(x)$ exists. Now it follows that $f_y(\varepsilon,y) < 0$, by (5.2). Hence by the implicit function theorem, we can solve for y as a function of x, the required $\phi(x)$ in a neighborhood of $x = \varepsilon$, and in fact the solution exists so long as f_y does not vanish (i.e., on $0 < x < B$). Further, since f_y is continuous, so is $\phi(x)$ [in fact, $\phi(x)$ is continuously differentiable]. The end points are included, since we know that $f(A,0) = f(0,B) = 0$ and that (5.11) precludes any other points on the axes from being included in the solution set of $f(x,y) = 0$. In Fig. 5.2, $y = \phi(x)$ is shown as curve C_f, the prey isocline.

In a similar manner, the existence of a function $x = \Psi(y)$ such that

$$g[\psi(y),y] = 0, \qquad \psi(0) = C, \qquad \lim_{y\to\infty} \psi(y) = \infty \tag{5.13}$$

can be shown. This curve, the predator isocline, is shown as curve C_g in Fig. 5.2. Clearly from the properties of $\phi(x)$ and $\psi(y)$ are from (5.10), C_f and C_g must intersect exactly once, giving the required equilibrium, which we label (x^*,y^*).

Now from hypothesis 11, $f(0,y) > 0$ for $y < A$ and $f(0,y) < 0$ for $y > A$. But for fixed y, $f(x,y)$ may not change sign by continuity, except at $y = \phi(x)$. Hence we find that

$$f(x,y) \lesssim 0, \quad \text{if } y \gtrless \phi(x) \tag{5.14}$$

Similarly, from hypothesis 12, we get

$$g(x,y) \lesssim 0, \quad \text{if } x \gtrless \psi(y)$$

Hence, from (5.1), this gives

$$\begin{aligned} x' &\lesssim 0, \quad \text{if } y \gtrless \phi(x) \\ y' &\lesssim 0, \quad \text{if } x \gtrless \psi(y) \end{aligned} \tag{5.15}$$

We now divide the first quadrant into the axes and four zones as shown in Fig. 5.2. The directions of the trajectories are as indicated there because of (5.15). We define one more curve C_k, where C_k is that curve such that

$$\frac{dy}{dx} = \frac{yg(x,y)}{xf(x,y)} = k \tag{5.16}$$

Choosing

$$k = \frac{g(B,0)}{Bf_y(B,0)} \tag{5.17}$$

we see that $k < 0$ and the curve passes through $(B,0)$. Further, using arguments similar to the first part of this proof, we see again that C_k with k given by (5.17) is a continuous curve representing x as a function of y, and since $k < 0$, it must lie to the right of C_g, and the range of y is $[0,\infty)$.

Now choose a point (x_0,y_0) initiating on C_k such that $y_0 > \sup_{0\leq x\leq B} \{y \mid y = \phi(x)\}$. Consider the solution Γ through (x_0,y_0). Then the behavior of Γ is exactly as the Γ of Sec. 4.6, and once more, by the Poincaré-Bendixon theorem, this theorem follows. For fuller details, see Albrecht et al. (1974).

5.4 WALTMAN'S MODIFICATION OF THE GROWTH EQUATIONS

In Waltman (1964), system (5.1) was modified into the form

$$\begin{aligned} x' &= \alpha x f(x,y) \\ y' &= y g(x,y) \end{aligned} \tag{5.18}$$

by the introduction of the parameter α. The purpose of the modification was to find a value of α, say α_0, so that in a one-sided neighborhood of α_0, or at α_0 itself, there would be periodic solutions of small amplitude. It will turn out that α_0 may be determined in terms of f and g under certain circumstances so as to obtain these small-amplitude periodic solutions. The main tool used is the Hopf bifurcation theorem (see Sec. F of the Appendix).

It is first supposed that there exists $a > 0$ and $b > 0$ such that

$$f(a,b) = g(a,b) = 0 \tag{5.19}$$

The bifurcation parameter ε is introduced by writing α as

$$\alpha = \alpha_0 + \varepsilon \tag{5.20}$$

where α_0, determined below, is such as to satisfy the first Hopf hypothesis. Then the matrix $A(\varepsilon)$ as described in Sec. F of the Appendix is

$$A(\varepsilon) = \begin{bmatrix} (\alpha_0 + \varepsilon) a f_x(a,b) & (\alpha_0 + \varepsilon) a f_y(a,b) \\ b g_x(a,b) & b g_y(a,b) \end{bmatrix} \tag{5.21}$$

Then the first Hopf condition

$$\mathrm{tr}\, A(0) = 0, \qquad \det A(0) > 0 \tag{5.22}$$

becomes

$$\alpha_0 a f_x(a,b) + b g_y(a,b) = 0 \tag{5.23}$$

$$\alpha_0 [f_x(a,b) g_y(a,b) - f_y(a,b) g_x(a,b)] > 0 \tag{5.24}$$

Condition (5.23), under the assumption $f_x(a,b) \neq 0$, leads to the choice of α_0,

$$\alpha_0 = \frac{-b g_y(a,b)}{a f_x(a,b)} \tag{5.25}$$

The matrix $A_\varepsilon(\varepsilon)$ is given by

$$A_\varepsilon(\varepsilon) = \begin{bmatrix} a f_x(a,b) & a f_y(a,b) \\ 0 & 0 \end{bmatrix} \tag{5.26}$$

and hence the second Hopf condition

$$\mathrm{tr}\, A_\varepsilon(0) \neq 0 \tag{5.27}$$

becomes

$$f_x(a,b) \neq 0 \tag{5.28}$$

which is already assumed. Hence if (5.24), (5.25), and (5.28) are valid, the required "small-amplitude" periodic solutions exist.

What is not given by the theorem is whether these periodic solutions occur for positive ε or negative ε, or at $\varepsilon = 0$ (the neutral case). If the periodic solutions are not neutral, it is known that for small $|\varepsilon|$ the amplitudes are of order $|\varepsilon|^{1/2}$ as $|\varepsilon| \to 0$.

Waltman's technique is utilized in Chap. 6 in discussing a model of immune response.

5.5 A GENERAL PERTURBED KOLMOGOROV-TYPE SYSTEM

The material in this and the succeeding two sections is taken from Freedman (1975). System (5.1) is now modified so as to include a general small parameter   as follows:

$$\begin{aligned} x' &= xf(x,y,\varepsilon) \\ y' &= yg(x,y,\varepsilon) \end{aligned} \tag{5.29}$$

As always, it is assumed that f and g are sufficiently smooth for whatever analysis we need. Further, we assume that there exists $a > 0$, $b > 0$ such that

$$f(a,b,0) = g(a,b,0) = 0 \tag{5.30}$$

We are interested in when there is a perturbed equilibrium and in its stability as a function of f, g, and their derivatives and ε.

For convenience of notation, let

$$\begin{aligned} \kappa_0 &= f_x(a,b,0), & \lambda_0 &= f_y(a,b,0), & \xi_0 &= f_\varepsilon(a,b,0) \\ \mu_0 &= g_x(a,b,0), & \nu_0 &= g_y(a,b,0), & \eta_0 &= g_\varepsilon(a,b,0) \end{aligned} \tag{5.31}$$

Further, let Δ be the matrix

$$\Delta = \begin{bmatrix} \kappa_0 & \lambda_0 \\ \mu_0 & \nu_0 \end{bmatrix} \tag{5.32}$$

and denote by $|\Delta|$ the determinant of Δ. We now assume for the remainder of this section that

$$|\Delta| \neq 0 \tag{5.33}$$

To establish the existence of a perturbed equilibrium, we must solve the algebraic system

$$f(x,y,\varepsilon) = g(x,y,\varepsilon) = 0 \tag{5.34}$$

near $x = a$, $y = b$, $\varepsilon = 0$ for x and y as functions of ε. By the implicit function theorem (see Sec. B.3 of the Appendix) since (5.33) holds, this can be done. Let $x^*(\varepsilon)$ and $y^*(\varepsilon)$ be this unique solution. Then from the implicit function theory they have the form

$$x^* = a + \frac{1}{|\Delta|}(\lambda_0\eta_0 - \nu_0\xi_0)\varepsilon + o(\varepsilon) = a + a_1\varepsilon + o(\varepsilon)$$
$$y^* = b + \frac{1}{|\Delta|}(\mu_0\xi_0 - \kappa_0\eta_0)\varepsilon + o(\varepsilon) = b + b_1\varepsilon + o(\varepsilon) \quad (5.35)$$

The stability of (x^*,y^*) is now given by the signs of the real parts of the eigenvalues of the variational matrix evaluated at $x = x^*$, $y = y^*$. Let $M(\varepsilon)$ be this variational matrix. Then

$$M(\varepsilon) = \begin{bmatrix} x^*(\varepsilon)f_x[x^*(\varepsilon),y^*(\varepsilon),\varepsilon] & x^*(\varepsilon)f_y[x^*(\varepsilon),y^*(\varepsilon),\varepsilon] \\ y^*(\varepsilon)g_x[x^*(\varepsilon),y^*(\varepsilon),\varepsilon] & y^*(\varepsilon)g_y[x^*(\varepsilon),y^*(\varepsilon),\varepsilon] \end{bmatrix} \quad (5.36)$$

We are interested in $M(\varepsilon)$ as a power series in ε. Hence, expanding f_x, f_y, g_x, and g_y in Taylor series in x^*, y^*, and ε and utilizing (5.35), we get

$$\begin{aligned} f_x(x^*,y^*,\varepsilon) &= a\kappa_0 + (a\kappa_1 + a_1\kappa_0)\varepsilon + o(\varepsilon) \\ f_y(x^*,y^*,\varepsilon) &= a\lambda_0 + (a\lambda_1 + a_1\lambda_0)\varepsilon + o(\varepsilon) \\ g_x(x^*,y^*,\varepsilon) &= b\mu_0 + (b\mu_1 + b_1\mu_0)\varepsilon + o(\varepsilon) \\ g_y(x^*,y^*,\varepsilon) &= b\nu_0 + (b\nu_1 + b_1\nu_0)\varepsilon + o(\varepsilon) \end{aligned} \quad (5.37)$$

where

$$\begin{aligned} \kappa_1 &= f_{xx}(a,b,0)a_1 + f_{xy}(a,b,0)b_1 + f_{x\varepsilon}(a,b,0) \\ \lambda_1 &= f_{xy}(a,b,0)a_1 + f_{yy}(a,b,0)b_1 + f_{y\varepsilon}(a,b,0) \\ \mu_1 &= g_{xx}(a,b,0)a_1 + g_{xy}(a,b,0)b_1 + g_{x\varepsilon}(a,b,0) \\ \nu_1 &= g_{xy}(a,b,0)a_1 + g_{yy}(a,b,0)b_1 + g_{y\varepsilon}(a,b,0) \end{aligned} \quad (5.38)$$

Define

$$\begin{aligned} p(\varepsilon) &= -[(a\kappa_0 + b\nu_0) + (a\kappa_1 + b\nu_1 + a_1\kappa_0 + b_1\nu_0)\varepsilon + o(\varepsilon)] \\ &= -[p_0 + p_1\varepsilon + o(\varepsilon)] \end{aligned} \quad (5.39)$$

and

$$\begin{aligned}D(\varepsilon) &= [(a\kappa_0 - b\nu_0)^2 + 4ab\lambda_0\mu_0] + [2(a\kappa_0 - b\nu_0)(a\kappa_1 + a_1\kappa_0 \\ &\quad - b\nu_1 - b_1\nu_0) + 4ab(\lambda_0\mu_1 + \lambda_1\mu_0) \\ &\quad + 4(ab_1 + a_1b)\lambda_0\mu_0]\varepsilon + o(\varepsilon) \\ &= D_0 + D_1\varepsilon + o(\varepsilon)\end{aligned} \tag{5.40}$$

Then the required eigenvalues are

$$\lambda = \frac{-p(\varepsilon)}{2} \pm \frac{\sqrt{D(\varepsilon)}}{2} \tag{5.41}$$

If Re $\lambda < 0$ for both eigenvalues, we have a stable equilibrium. If Re $\lambda > 0$, the equilibrium is unstable. If the eigenvalues are such that one has a positive real part and the other has a negative real part, the equilibrium is said to be hyperbolic.

We now analyze the equilibrium for sufficiently small ε up to the order of ε.

1. Suppose $D_0 < 0$ and $p_0 < 0$. Then the equilibrium is a stable spiral point.

Note that $p_0 < 0$ is satisfied if $\kappa_0, \nu_0 < 0$, which corresponds to the case of diminishing returns of Samuelson (1967).

2. Suppose $D_0 < 0$, $p_0 > 0$. Then the equilibrium is an unstable spiral point.

Note that $p_0 > 0$ is satisfied if $\kappa_0, \nu_0 > 0$, which corresponds to the case of increasing returns of Samuelson (1967).

3. Suppose $D_0 < 0$, $p_0 = 0$, $p_1 < 0$. Then the equilibrium is a stable spiral.

Note that $D_0 < 0$, $p_0 = 0$ is satisfied by the Lotka-Volterra model.

4. Suppose $D_0 < 0$, $p_0 = 0$, $p_1 > 0$. Then the equilibrium is an unstable spiral.
5. Suppose $D_0 = 0$, $D_1 < 0$, $p_0 < 0$. Then the equilibrium is a stable spiral.

6. Suppose $D_0 = 0$, $D_1 < 0$, $p_0 > 0$. Then the equilibrium is an unstable spiral.

Note that D_0 can be written

$$D_0 = p_0^2 - 4ab|\Delta| \tag{5.42}$$

Hence if $p_0 = 0$, then $D_0 = -4ab|\Delta|$, and so $D_0 = 0$, $p_0 = 0$ is impossible.

7. Suppose $D_0 = 0$, $D_1 > 0$, $p_0 < 0$. Then the equilibrium is a stable node.
8. Suppose $D_0 = 0$, $D_1 > 0$, $p_0 > 0$. Then the equilibrium is an unstable node.
9. Suppose $D_0 > 0$, $\Delta > 0$, $p_0 < 0$. Then the equilibrium is a stable node.
10. Suppose $D_0 > 0$, $\Delta > 0$, $p_0 > 0$. Then the equilibrium is an unstable node.

Note that $D_0 > 0$, $\Delta > 0$, $p_0 = 0$ by (5.42) is impossible.

11. Suppose $D_0 > 0$, $\Delta < 0$. Then for any value of p_0, the equilibrium is hyperbolic.

We do not consider here cases where $p_0 = p_1 = 0$ or $D_0 = D_1 = 0$. We define the sets A_1, B_1, C_1 as follows:

$$\begin{aligned} A_1 &= \{1, 3, 5, 7, 9\} \\ B_1 &= \{2, 4, 6, 8, 10\} \\ C_1 &= \{11\} \end{aligned} \tag{5.43}$$

By the preceding, we have proved the following:

THEOREM 5.2 *Let* $f(x,y,\varepsilon)$, $g(x,y,\varepsilon)$ *be such that the hypotheses of* k *hold. Then if* $k \in A_1$, (x^*,y^*) *is a stable equilibrium. If* $k \in B$ (x^*,y^*) *is an unstable equilibrium. If* $k \in C_1$, (x^*,y^*) *is a hyperbolic equilibrium.*

In the next section several critical cases are examined.

5.6 CRITICAL CASES

In this section we are interested in investigating the case where

$$|\Delta| = 0 \tag{5.44}$$

where Δ is given by (5.32). Without loss of generality it may be assumed that Δ is in Jordan canonical form (see Coddington and Levinson, 1955, Chap. 3). This can always be effected anyway by an appropriate rotation and stretching of axes, which does not change the stability of the equilibrium. Since $|\Delta| = 0$, Δ must be one of three distinct types.

$$\text{I:} \quad \Delta = \begin{bmatrix} 0 & 0 \\ 0 & \nu_0 \end{bmatrix}, \qquad \nu_0 \neq 0$$

$$\text{II:} \quad \Delta = \begin{bmatrix} 0 & 1 \\ 0 & 0 \end{bmatrix}$$

$$\text{III:} \quad \Delta = \begin{bmatrix} 0 & 0 \\ 0 & 0 \end{bmatrix}$$

We will discuss each of these separately.

We suppose first that

$$\kappa_0 = \lambda_0 = \mu_0 = 0, \qquad \nu_0 \neq 0 \tag{5.45}$$

We are again interested in whether or not an equilibrium close to (a,b) exists for the system (5.29) for small ε. Since $\nu_0 \neq 0$, $g(x,y,\varepsilon) = 0$ can be solved for y as a function of x and ε, giving

$$y = b - \frac{\eta_0}{\nu_0}\varepsilon + \text{HOT} \tag{5.46}$$

We substitute (5.46) into f and define

$$G(x,\varepsilon) = f[x,y(x,\varepsilon),\varepsilon] \tag{5.47}$$

Now $G(a,0) = G_x(a,0) = 0$. Computing $G_\varepsilon(a,0)$, we get

$$G_\varepsilon(a,0) = \xi_0 \tag{5.48}$$

Our analysis now depends on whether $\xi_0 = 0$ or $\xi_0 \neq 0$.

We now suppose that $\xi_0 \neq 0$. We now compute $G_{xx}(a,0)$, which after simplification becomes

$$G_{xx}(a,0) = f_{xx}(a,b,0) \tag{5.49}$$

Hence, by the implicit function theorem (Sec. B.2 of the Appendix), if

$$f_{xx}(a,b,0)\xi_0 < 0 \quad (> 0) \tag{5.50}$$

we can solve $G(x,\varepsilon) = 0$ for x as a function of ε for sufficiently small positive (negative) ε in the form

$$x = a + \theta_0\varepsilon^{1/2} + o(\varepsilon^{1/2})$$
$$\left\{x = a + \theta_0(-\varepsilon)^{1/2} + o[(-\varepsilon)^{1/2}]\right\} \tag{5.51}$$

where θ_0 is either of

$$\theta_0 = \pm\left[-\frac{2\xi_0}{f_{xx}(a,b,0)}\right]^{1/2} \qquad \left\{\theta_0 = \pm\left[\frac{2\xi_0}{f_{xx}(a,b,0)}\right]^{1/2}\right\} \tag{5.52}$$

Hence there are two branches of equilibria emanating from (a,b) in the case of small positive ε,

$$x^* = a + \theta_0\varepsilon^{1/2} + o(\varepsilon^{1/2}), \qquad y^* = b + o(\varepsilon^{1/2}) \tag{5.53}$$

with similar expressions in the case of negative ε.

It is once more possible to analyze the stability of these branches, by linearizing about (x^*,y^*). If the analysis is carried through, we arrive at the following possiblities (in the case of positive ε):

12. Suppose $\nu_0 < 0$, $\theta_0 f_{xx}(a,b,0) < 0$. Then the equilibrium is a stable node.
13. Suppose $\nu_0 < 0$, $\theta_0 f_{xx}(a,b,0) > 0$. Then the equilibrium is hyperbolic.
14. Suppose $\nu_0 > 0$, $\theta_0 f_{xx}(a,b,0) < 0$. Then the equilibrium is hyperbolic.

15. Suppose $\nu_0 > 0$, $\theta_0 f_{xx}(a,b,0) > 0$. Then the equilibrium is an unstable node.

If now the assumption is made that $\xi_0 = 0$, then $G(a,0) = 0$. By similar analysis to the preceding, the stability of the perturbed equilibrium can again be determined. First if $G_{xx}(a,0) = 0$, then

16. Suppose $f_{xx}(a,b,0) = 0$, $\nu_0 < 0$, $f_{x\varepsilon}(a,b,0) - f_{xy}(a,b,0) \times \eta_0/\nu_0 < 0$. Then the equilibrium is a stable node.
17. Suppose $f_{xx}(a,b,0) = 0$, $\nu_0 < 0$, $f_{x\varepsilon}(a,b,0) - f_{xy}(a,b,0) \times \eta_0/\nu_0 > 0$. Then the equilibrium is hyperbolic.
18. Suppose $f_{xx}(a,b,0) = 0$, $\nu_0 > 0$, $f_{x\varepsilon}(a,b,0) - f_{xy}(a,b,0) \times \eta_0/\nu_0 < 0$. Then the equilibrium is hyperbolic.
19. Suppose $f_{xx}(a,b,0) = 0$, $\nu_0 > 0$, $f_{x\varepsilon}(a,b,0) - f_{xy}(a,b,0) \times \eta_0/\nu_0 > 0$. Then the equilibrium is an unstable node.

We now assume that $f_{xx}(a,b,0) \neq 0$. Then we get

20. Suppose $\nu_0 < 0$ and α_0 is such that $\lambda_{21} < 0$. Then the equilibrium is a stable node.
21. Suppose $\nu_0 < 0$ and α_0 is such that $\lambda_{21} > 0$. Then the equilibrium is hyperbolic.
22. Suppose that $\nu_0 > 0$ and α_0 is such that $\lambda_{21} < 0$. Then the equilibrium is hyperbolic.
23. Suppose that $\nu_0 > 0$ and α_0 is such that $\lambda_{21} > 0$. Then the equilibrium is an unstable node.

Here λ_{21} is given by the expression

$$\lambda_{21} = \pm a[G_{x\varepsilon}(a,0)^2 - G_{xx}(a,0)G_{\varepsilon\varepsilon}(a,0)]^{1/2} \tag{5.54}$$

From the preceding considerations, when Jordan form I occurs, we see that if $\xi_0 = f_{xx}(a,b,0) = 0$, there will be one branch of equilibria originating from (a,b). The equilibria on this branch will all be stable or unstable or hyperbolic. Otherwise there will be two branches originating from (a,b). The equilibria on one of these branches will all be either stable or unstable, and the equilibria on the other branch will all be hyperbolic.

The analysis when Δ has Jordan form II is similar to the preceding analysis. In this case if $\eta_0 g_{xx}(a,b,0) = 0$, there is one branch of equilibria which may be stable, unstable, or hyperbolic. Otherwise there will be two branches, on one of which the equilibria are either stable or unstable and on the other of which they are hyperbolic.

If the Jordan form of Δ is of type III, the analysis is more complicated, but still may be carried through. It turns out that if $\xi_0^2 + \eta_0^2 \neq 0$, there will be either two or four branches of equilibria. In pairs they must be either both hyperbolic or one stable and one unstable. If, however, $\xi_0 = \eta_0 = 0$, there may be one, two, three, or four branches with any combination of stability occurring.

The reader is referred to Freedman (1975) for complete details of the analysis.

5.7 THE EXISTENCE OF LIMIT CYCLES FOR THE PERTURBED SYSTEM

In a manner analagous to the techniques of Sec. 3.6, several criteria for the existence of a limit cycle about the perturbed equilibrium can be stated. Basically the idea is again to use the Poincare-Bendixon theorem.

Define

$$V(x,y) = \frac{1}{2}[y^* g_x(x^*,y^*,\varepsilon)(x - x^*)^2 + x^* f_y(x^*,y^*,\varepsilon)(y - y^*)^2] \tag{5.55}$$

$$\begin{aligned} W(x,y) = {} & x^* y^* [f_x(x^*,y^*,\varepsilon) g_x(x^*,y^*,\varepsilon)(x - x^*)^2 \\ & - f_y(x^*,y^*,\varepsilon) g_y(x^*,y^*,\varepsilon)(y - y^*)^2] \\ & + y^* g_x(x^*,y^*,\varepsilon)(x - x^*)\tilde{f}(x,y,\varepsilon) \\ & - x^* f_y(x^*,y^*,\varepsilon)(y - y^*)\tilde{g}(x,y,\varepsilon) \end{aligned} \tag{5.56}$$

where $\tilde{f}$ and $\tilde{g}$ are terms of higher order than quadratic of f and g in expansions about x^* and y^*. Let $g_x(x^*,y^*,\varepsilon) > 0$ and

$f_y(x^*,y^*,\varepsilon) < 0$ for sufficiently small positive ε. Let Γ_k be the closed curve defined by $V(x,y) = k$. Suppose that $f(x,y,\varepsilon)$ and $g(x,y,\varepsilon)$ are such that (1) (x^*,y^*) is unstable, and (2) there exists k such that Γ_k lies in the first quadrant and $W(x,y) < 0$ for $(x,y) \in \Gamma_k$. Then since a direct computation gives that

$$\dot{V}(x,y) = W(x,y) \tag{5.57}$$

by the Poincaré-Bendixon theorem there must be a stable limit cycle.

A second criterion, taking the hint from Sec. 3.6, may be obtained by taking

$$V(x,y) = \Phi(x) + \Psi(y)$$

where the $\Phi(x)$ and $\Psi(y)$ are obtained by integrating the terms corresponding to the Lotka-Volterra equation. Then, again $\dot{V} < 0$ is the criterion required.

NOTES ON THE LITERATURE

Conditions under which the general Kolmogorov model behaves like a predator-prey system near equilibrium were first given by Brauer (1972). Kolmogorov-type models involving constant harvesting have been discussed by Brauer and Sanchez (1975a,b). Reddingius (1963) has considered a Kolmogorov-type model of consumer-food relationships. Albrecht et al. (1976) have considered a Kolmogorov model with controls. Rapport and Turner (1975) have considered a general interaction model.

EXERCISES

5.1 Carry out the integrations and differentiations suggested at the end of Sec. 5.7 to obtain a criterion for a stable limit cycle.

5.2 (Difficult but can be solved) For the model described by (5.18), derive a criterion which guarantees that at $\alpha = \alpha_0$, the case of neutral stability does not occur (i.e., that there is a nondegenerate Hopf bifurcation.

5.3 (Open problem) Analyze a Kolmogorov-type model of the form

$$x' = xf(t,x,y,\varepsilon), \qquad y' = yg(t,x,y,\varepsilon)$$

where f and g are periodic in t.

5.4 (Open problem) Derive criteria under which limit cycles in Kolmogorov-type models are unique.

Chapter 6
RELATED TOPICS AND APPLICATIONS

A: RELATED TOPICS

Two-species predator-prey systems may be extended in several directions which relax the "two-species" requirement. One extension would be to consider predator-prey systems involving more than two trophic levels. Here the simplest such extension will be considered, namely, three-species food chains. A second direction would be to extend the number of species in two trophic levels. Again the simplest such extension is considered here, namely, the question of one predator and two prey. The question of two predators feeding on one prey properly belongs with the competition topics and is discussed in Part III.

The work on food chains and part of the work on one-predator, two-prey systems is based on work by the author and Paul Waltman.

6.1 SIMPLE FOOD CHAINS: A GENERAL MODEL

By a simple food chain is meant a line of succession of prey and predators, where each predator feeds only on the prey directly lower than itself in trophic level and, except for the highest predator, is fed on only by the predator directly higher than itself in trophic level. We then have a situation of A eats B eats C eats ... eats K. Here we consider only a three-species food chain. A eats B eats C. The model utilized will be a generalization to three species of the continuous intermediate model discussed in Chap. 4. The material in this and the next section is based on Freedman and Waltman (1977b).

In what follows, by persistence of a species is meant the continued existence in the deterministic sense, i.e., $\limsup_{t\to\infty} N(t) > 0$, where N(t) is the population of species N at time t.

We consider now the following system as a model simulating a food chain, where x is the number of lowest trophic species or prey, y is the number of middle trophic level species or first predator, and z is the number of highest trophic level species or second predator:

$$\begin{aligned} x' &= xg(x) - yp(x) \\ y' &= y[-r + cp(x)] - zq(y) \qquad (' = \tfrac{d}{dt}) \\ z' &= z[-s + dq(y)] \end{aligned} \tag{6.1}$$

where r, s, c, and d are positive constants.

Here g(x) is the specific growth rate of the prey and is always assumed to satisfy

(H1) $g(0) = \alpha > 0, \qquad g_x(x) \leq 0$ for $x \geq 0$

In those models in which the assumption is made that the environment has a natural carrying capacity, we will also assume

(H2) $\exists K > 0 \ni g(K) = 0$

The first predator functional response p(x) is assumed to satisfy

(H3) $p(0) = 0, \qquad p_x(x) > 0$ for $x \geq 0$

Similarly, q(y), the second predator functional response, is assumed to satisfy

(H4) $q(0) = 0, \qquad q_y(y) > 0$ for $y \geq 0$

We note that under (H3) and (H4) the predation curves include the usual curves found in the literature as described in Chap. 2.

Now the question of existence of equilibria for system (6.1) is considered. First note that

(E1) (0,0,0)

is always an equilibrium. Further, if (H2) is satisfied, then clearly

(E2) (K,0,0)

is also an equilibrium.

We now want conditions that guarantee an equilibrium in the interior of the first quadrant of the x-y plane. This has already been done in Chap. 4, and translated into the present variables. These conditions become, using ^ instead of *,

$$\frac{r}{c} \in \text{range } [p(x)] \tag{6.2}$$

$$p(\hat{x}) = \frac{r}{c} \tag{6.3}$$

defining $\hat{x}$.

$$\hat{y} = \frac{\hat{x}g(\hat{x})}{p(\hat{x})} \tag{6.4}$$

defining $\hat{y}$ and

$$\hat{x} < K \tag{6.5}$$

Then in case such $\hat{x}$ and $\hat{y}$ exist,

(E3) $(\hat{x},\hat{y},0)$ is an equilibrium

The question of whether or not there exists an equilibrium in the interior of the first octant is also of interest. If such exists it will be labeled

(E4) (x^*,y^*,z^*)

The condition for y^* to exist is clear from the third part of Eq. (6.1).

$$\frac{s}{d} \in \text{range } [q(y)] \tag{6.6}$$

Then there is a y^* such that

$$q(y^*) = \frac{s}{d} \tag{6.7}$$

From the first part of Eq. (6.1) we can solve for x^* in terms of y^* provided that

$$y^* \in \text{range}\left[\frac{xg(x)}{p(x)}\right], \qquad x \geq 0 \tag{6.8}$$

in which case there may be more than one such x^*. From the second of Eq. (6.1), z^* is given by

$$z^* = \frac{y^*[-r + cp(x^*)]}{q(y^*)} \tag{6.9}$$

Here z^* is positive, provided that $-r + cp(x^*) > 0$, or that

$$x^* > \hat{x} \tag{6.10}$$

Hence if (6.6), (6.8), and (6.10) are satisfied, (E4) exists.

To compute the stability of the equilibria we need the respective variational matrices. Let $V(x,y,z)$ denote the variational matrix of (6.1) for general x, y, z. Then

$$V(x,y,z) = \begin{bmatrix} xg_x(x) + g(x) - yp_x(x) & -p(x) & 0 \\ cyp_x(x) & -r + cp(x) - zq_y(y) & -q(y) \\ 0 & dzq_y(y) & -s + dq(y) \end{bmatrix} \tag{6.11}$$

In Chap. 4, the equilibria (E1), (E2), and (E3) were examined for their stability in the x-y plane. Here we need to examine stability in the z direction as well. Letting V_i, $i = 1,\ldots,4$, denote the value of $V(x,y,z)$ at (Ei) gives

$$V_1 = \begin{bmatrix} \alpha & 0 & 0 \\ 0 & -r & 0 \\ 0 & 0 & -s \end{bmatrix}, \qquad V_2 = \begin{bmatrix} Kg_x(K) & -p(K) & 0 \\ 0 & -r + cp(K) & 0 \\ 0 & 0 & -s \end{bmatrix} \tag{6.12}$$

Clearly, then, from (6.12) near (E1) the prey population grows while both predator populations decline, and near (E2) the first predator population grows and the second predator population declines. Note the saddle-point nature of both.

We now concern ourselves with the equilibrium (E3).

$$V_3 = \begin{bmatrix} \hat{x}g_x(\hat{x}) + g(\hat{x}) - \hat{y}p_x(\hat{x}) & -p(\hat{x}) & 0 \\ c\hat{y}p_x(\hat{x}) & 0 & -q(\hat{y}) \\ 0 & 0 & -s + dq(\hat{y}) \end{bmatrix} \tag{6.13}$$

The eigenvalue governing the stability in the z direction is $-s + dq(\hat{y})$. We know that $\hat{x} < x^*$ if x^* exists, but the relation between $\hat{y}$ and y^* is not in general determined. Hence $-s + dq(\hat{y})$ may be positive, negative, or zero. If $-s + dq(\hat{y}) < 0$ and if there are also no periodic orbits in the x-y plane, then there are orbits in the interior of the first octant that approach (E3).

Suppose now that $-s + dq(\hat{y}) > 0$ and that there are no nontrivial periodic orbits in the open positive quadrant of the x-y plane; since all the equilibria in the x-y plane are saddle points, no orbit in the interior of the first octant can approach the x-y plane as $t \to \infty$. By the preceding we have proved the following theorem.

THEOREM 6.1 *Let* (6.1) *be such that there are no nontrivial periodic solutions in the* x-y *plane. Then a necessary condition for the persistence of all three species for arbitrary positive initial populations is*

$$-s + dq(\hat{y}) \geq 0 \tag{6.14}$$

and a sufficient condition for the persistence of all three species for arbitrary positive inital populations is

$$-s + dq(\hat{y}) > 0 \tag{6.15}$$

A similar theorem can be proved in the case that there are nontrivial periodic solutions in the x-y plane as well. This is given in Theorem 6.2.

THEOREM 6.2 *Let* (E3) *be an unstable equilibrium in the* x *and* y *directions or let* (6.15) *hold. Further, for each periodic solution*

$x = \phi(t)$, $y = \psi(t)$ *in the first quadrant of the plane which is stable in the plane on at least one side, let*

$$-s + \frac{d}{T}\int_0^T q[\psi(t)]\, dt > 0 \tag{6.16}$$

Then all three species persist for all time.

Now it will be shown that if (H2) is satisfied, then all populations are bounded whether or not (E4) exists. Then in the case that (E4) exists, its stability will be examined.

Suppose that (H2) does indeed hold. Then by the well-known property of logistic growth, the prey is limited by its carrying capacity. Specifically, given initial values of system (6.1), (x_0, y_0, z_0), then from the first of Eq. (6.1)

$$x' \leq xg(x) \tag{6.17}$$

and by the usual comparison theorem, we have that

$$x(t) \leq \ell, \qquad \text{where } \ell = \max(x_0, K) \tag{6.18}$$

If we add c times the first equation of (6.1) to the second, we get $(cx + y)' = cxg(x) - ry - zg(y) \leq cxg(0) - ry = -cg(0)x - ry + 2cg(0)x \leq -m_1[cxg(0) + y] + 2cg(0)\ell$, where $m_1 = \min[g(0), r]$. Hence

$$cx + y \leq (cx_0 + y_0)e^{-m_1 t} + \frac{2cg(0)\ell}{m_1}$$

Setting $\bar{y} = cx_0 + y_0 + 2cg(0)\ell/m_1$, we get $y(t) \leq \bar{y}$. Now we add d times the second part of Eq. (6.1) to the third and obtain

$$(dy + z)' = dry - sz + cdp(x)y \tag{6.19}$$

Let $m = \min(r,s)$. Then $-dry - sz \leq -m(dy + z)$, and using (6.19) with $w = dy + z$, we obtain

$$w' \leq -mw + cdp(\ell)\bar{y} \tag{6.20}$$

which implies

$$w \leq w_0 e^{-mt} + \frac{cdp(\ell)\bar{y}}{m} \tag{6.21}$$

or, using the standard comparison theorem,

$$0 \leq dy(t) + z(t) \leq (dy_0 + z_0)e^{-mt} + \frac{cdp(t)\bar{y}}{m} \tag{6.22}$$

Thus z(t) is also bounded. The following theorem has been established.

THEOREM 6.3 *Let hypotheses* (H1) *and* (H2) *hold. Let* $p(x) > 0$ *for* $x > 0$. *Then all solutions of system* (6.1) *initiating in the first octant are bounded.*

Note that this theorem agrees with biological intuition. If the prey species is resource limited, then both predator species are also limited, regardless of their predation curve shapes.

We can state a consequence of this in the case the hypotheses of Theorems 6.1 or 6.2 hold.

COROLLARY 6.4 *Let the hypotheses of Theorem* 6.3 *and either Theorem* 6.1 *or Theorem* 6.2 *hold. Then there exists a recurrent motion lying in the first octant.*

PROOF. Since all trajectories initiating in the first octant are bounded and lie in that octant, by a well-known theorem of dynamical systems (see Nemytskii and Stepanov, 1960) the corollary is proved.

We now suppose (H1) to (H4) and that (E4) exists. Then

$$V_4 = \begin{bmatrix} m_{11} & m_{12} & 0 \\ m_{21} & m_{22} & m_{23} \\ 0 & m_{32} & 0 \end{bmatrix}$$

where

$$m_{11} = x^* g_x(x^*) + g(x^*) - y^* p_x(x^*), \qquad m_{12} = -p(x^*) < 0$$

$$m_{21} = cy^* p_x(x^*) > 0, \qquad m_{22} = -r + cp(x^*) - z^* q_y(y^*) \tag{6.23}$$

$$m_{23} = -q(y^*) < 0, \qquad m_{32} = dz^* q_y(y^*) > 0$$

The characteristic polynomial whose roots are the eigenvalues of V_4 is then

$$f(\lambda) = \lambda^3 - (m_{11} + m_{22})\lambda^2 + (m_{11}m_{22} - m_{12}m_{21} - m_{23}m_{32}) + m_{11}m_{23}m_{32} \tag{6.24}$$

Since

$$f(0) = m_{11}m_{23}m_{32}, \qquad f(m_{11}) = -m_{11}m_{12}m_{21} \tag{6.25}$$

and since $m_{23}m_{32} < 0$ and $m_{12}m_{21} < 0$, either

$$m_{11} = 0 \qquad \text{or} \qquad f(0)f(m_{11}) < 0 \tag{6.26}$$

Hence there is a real root either at 0 or between 0 and m_{11}. As a consequence we see that if $m_{11} > 0$, (E4) is unstable.

Suppose now that $m_{11} < 0$. Let $\rho < 0$, $0 < |\rho| < |m_{11}|$, be the negative real root deduced above. Then, on dividing $f(\lambda)$ by $\lambda - \rho$, we obtain the quadratic

$$f_1(\lambda) = \lambda^2 + (\rho - m_{11} - m_{22})\lambda + [m_{11}m_{22} - m_{12}m_{21} - m_{23}m_{32} + \rho^2 - (m_{11} + m_{22})] \tag{6.27}$$

the roots of which are the remaining two eigenvalues. Since $\rho - m_{11} > 0$, if $m_{22} \leq 0$, then

$$\rho - m_{11} - m_{22} > 0 \tag{6.28}$$

which implies that the roots of $f_1(\lambda)$ have negative real parts. Hence we have proved the following theorem.

THEOREM 6.5 *Let* (H1) *to* (H4) *hold and suppose* (E4) *exists. If* $m_{11} > 0$, *then* (E4) *is unstable. If* $m_{11} \leq 0$ *and* $m_{22} \leq 0$, *then* (E4) *is stable. If further* $m_{11} < 0$, *then* (E4) *is asymptotically stable.*

Examining m_{11} and m_{22} in a little more detail, from (6.23) and the definition of x^*,

$$m_{11} = x^*g(x^*) \frac{d}{dz} \ln \left[\frac{xg(x)}{p(x)}\right]_{x=x^*}$$

Hence $m_{11} < 0$ (>0) if and only if $xg(x)/p(x)$ is decreasing (increasing) at x^*. Similarly,

$$m_{22} = [-r + cp(x^*)] \left[1 - \frac{y^*q_y(y^*)}{q(y^*)}\right]$$

or

$$m_{22} = [-r + cp(x^*)]y^* \frac{d}{dy} \ln \left[\frac{y}{q(y)}\right]\Bigg|_{y=y^*} \tag{6.29}$$

Since $x^* > \hat{x}$, and so $-r + cp(x^*) > 0$, then $m_{22} < 0$ (>0) if and only if $y/q(y)$ is decreasing (increasing) at y^*. The condition for m_{11} is related to the conditions discussed in Chap. 4.

We note that if $q(y)$ is a Holling-type predation, then $m_{22} > 0$. However, if $q(y)$ is linear, as in the Lotka-Volterra case, then $m_{22} = 0$. For predations curves with learning effects such as those shown in Haynes and Sisojevic (1966) m_{22} will be negative.

6.2 SIMPLE FOOD CHAINS: EXAMPLES

EXAMPLE A. In this example we assume that the functions g, p, q yield the Lotka-Volterra dynamics. More specifically, we consider the system

$$\begin{aligned} x' &= a_1x - a_{12}xy \\ y' &= -a_2y + a_{21}xy - a_{23}yz \\ z' &= -a_3z + a_{32}yz \end{aligned} \tag{6.30}$$

$$x(0) = \alpha_1 > 0, \qquad y(0) = \alpha_2 > 0, \qquad z(0) = \alpha_3 > 0$$

where all the constants are positive. Note that this system does not have a natural carrying capacity, i.e., (H2) is not satisfied. In this case the analysis of the preceding section may be completed to yield an exact answer to the question of the persistence of all three species.

THEOREM 6.6 *A necessary and sufficient condition for the persistence of all three species in a dynamical system governed by the system* (6.30) *is that* $\mu = a_1a_{32} - a_3a_{12} \geq 0$.

The details of the proof may be found in Freedman and Waltman (1977b). In the case that $\mu = 0$, it is shown there that all solutions are periodic.

EXAMPLE B. We modify the model of the previous example by introducing a carrying capacity into the dynamics of the lowest trophic level. Specifically, we consider the system

$$\begin{aligned} x' &= x\left[a_1\left(1 - \frac{x}{K}\right) - a_{12}y\right] \\ y' &= y(-a_2 + a_{21}x - a_{23}z) \\ z' &= z(-a_3 + a_{32}y) \end{aligned} \tag{6.31}$$
$$x(0) = \alpha_1, \qquad y(0) = \alpha_2, \qquad z(0) = \alpha_3$$

where all the constants are positive. The carrying capacity is K. (H1) to (H4) are satisfied and from Theorem 6.3 we know that all solutions with the preceding initial conditions are bounded (i.e., the closure of any trajectory is compact). We first analyze the equilibria. As noted in Sec. 6.1, (E1) and (E2) exist and are hyperbolic. The interest then focuses on (E3) and (E4).

For the system (6.51) (E3) is given by

$$(\hat{x},\hat{y},0) = \left[\frac{a_2}{a_{21}}, \frac{a_1(a_{21}K - a_2)}{a_{12}a_{21}K}, 0\right] \tag{6.32}$$

where for $\hat{y}$ to be positive we must assume

$$K > \frac{a_2}{a_{21}} \tag{6.33}$$

The variational matrix (6.13) takes the form

$$V_3 = \begin{bmatrix} \frac{-a_1}{K}\hat{x} & -a_{12}\hat{x} & 0 \\ a_{21}\hat{y} & 0 & -a_{23}\hat{y} \\ 0 & 0 & -a_3 + a_{32}\hat{y} \end{bmatrix} \tag{6.34}$$

Viewed as an equilibrium in the x-y plane (x,y) is asumptotically stable, since $(-a_1/K)\hat{x} < 0$. Further, the equilibrium will be unstable in the z direction if $-a_3 + a_{32}\hat{y} > 0$.

$$-a_3 + a_{32}\hat{y} = \frac{a_{32}a_1(a_{21}K - a_2) - a_3a_{21}a_{12}K}{a_{12}a_{21}K} \tag{6.35}$$

This quantity will be positive if

$$a_{32}a_1 - a_3a_{12} > 0$$

and (6.36)

$$K > \frac{a_2a_{32}a_1}{a_{21}(a_{32}a_1 - a_3a_{12})}$$

If $a_{32}a_1 - a_3a_{12} \leq 0$ or if K strictly violates (6.36), the equilibrium (x,y,0) will be asymptotically stable.

If $z(0) = 0$, the remaining two-dimensional system (6.31) may have limit cycles, at least one of which must be semistable. For these periodic solutions we are unable to obtain any more specific information than is already given in Theorem 6.2.

We consider now the existence of (E4). Solving for the equilibrium, one obtains

$$x^* = \frac{(a_1a_{32} - a_{12}a_3)K}{a_1a_{32}}$$

$$y^* = \frac{a_3}{a_{32}} \tag{6.37}$$

$$z^* = \frac{a_{21}(a_1a_{32} - a_{12}a_3)K - a_1a_2a_{32}}{a_1a_{23}a_{32}}$$

and to be interior to the positive octant requires precisely (6.36).

The variational matrix (V4) has entries

$$\begin{aligned} m_{11} &= \frac{-a_1x^*}{K} & m_{12} &= -a_{12}x^* \\ m_{21} &= a_{21}y^* & m_{22} &= 0 \\ m_{23} &= -a_{23}y^* & m_{32} &= a_{32}z^* \end{aligned} \tag{6.38}$$

With the preceding conditions making x^*, y^*, z^* positive, m_{11} 0 and $m_{22} = 0$, Theorem 6.5 implies that (E4) is asymptotically stable.

Finally we note that the asymptotic stability criteria are local and not global. The possibility of limit cycles in the plane exists as well as the possibility of more general three-dimensional limit sets (necessarily containing recurrent solutions). The analysis of the stability of such sets appears to be a very difficult problem.

EXAMPLE C. We modify the model of the previous example to include a Holling-type predation of the first predator on the prey. Specifically, we consider the system

$$\begin{aligned} x' &= x\alpha(1 - \frac{x}{K}) - \frac{\beta xy}{1 + ax} \\ y' &= y(-r + \frac{c\beta x}{1 + ax} - \gamma z) \\ z' &= z(-s + d\gamma y) \\ x(0) &= \alpha_1, \qquad y(0) = \alpha_2, \qquad z(0) = \alpha_3 \end{aligned} \tag{6.39}$$

The equilibrium (E3) is given by

$$\begin{aligned} \hat{x} &= \frac{r}{c\beta - ra} \\ \hat{y} &= \frac{\alpha c[K(c\beta - ra) - r]}{K(c\beta - ra)^2} \\ \hat{z} &= 0 \end{aligned} \tag{6.40}$$

For $\hat{x}$ and $\hat{y}$ to be positive requires

$$c\beta - ra > 0$$
$$K > \frac{r}{c\beta - ra} \tag{6.41}$$

Utilizing known criteria for stability of the two-dimensional system as given in Chap. 4, the conditions

$$K < \frac{1}{a} + \frac{2r}{c\beta - ra} \tag{6.42}$$

with $c\beta - ra > 0$ and

$$\frac{a}{c\beta - ra} < K < \frac{1}{a} + \frac{2r}{c\beta - ra} \tag{6.43}$$

guarantee that $(\hat{x},\hat{y},0)$ exists with $\hat{x} > 0$, $\hat{y} > 0$, and is asymptotically stable in the x-y plane. S. B. Hsu (1976) using a theorem of Dulac (see Andronov et al, 1973, p. 205) has shown that in this case the two-dimensional system has no limit cycles. Since there are no other equilibria in the open positive quadrant and since all solutions are bounded, the absence of limit cycles and the Poincare-Bendixson theorem (see Sec. E of the Appendix) allows one to conclude that a solution of (6.39) with $\alpha_3 = 0$ satisfies

$$\lim_{t\to\infty} x(t) = \hat{x}$$
$$\lim_{t\to\infty} y(t) = \hat{y} \tag{6.44}$$

If $K > 1/a + 2ar/(c\beta - ra)$, S. B. Hsu (1976), using Albrecht et al. (1974), has shown that there exists at least one periodic orbit (outermost, semistable, outside; innermost, semistable, inside).

Assuming (6.43) holds (and, of course, $c\beta - ra > 0$), then Theorem 6.1 says *that all species will persist if* $\hat{y} > s/dr$.

For the equilibrium (E4) we note first that

$$y^* = \frac{s}{d\gamma} > 0 \tag{6.45}$$

x^* is given as a root of

$$a\alpha x^2 - x(aK - 1)\alpha - K(\alpha - \beta y^*) = 0 \tag{6.46}$$

For there to be a positive root either

$$y^* < \alpha/\beta \tag{6.47}$$

or

$$\begin{aligned} y^* &\geq \alpha/\beta \\ K &> 1/a \end{aligned} \tag{6.48}$$

is required. In the second case there are two equilibria x_1^*, x_2^*. The final component is

$$z^* = \frac{(c\beta - ra)x^* - r}{\gamma(1 + ax^*)} \tag{6.49}$$

which is positive if

$$x^* > \frac{r}{c\beta - ra} = \hat{x} \tag{6.50}$$

Suppose there is an interior equilibrium. In the notation of (6.27)

$$m_{11} = x^*\left[-\frac{\alpha}{K} + \frac{a\beta y}{(1 + ax^*)^2}\right] < 0 \tag{6.51}$$

and by (6.27) $m_{22} = 0$ since $4/g(y) = \gamma$. Thus (E4) is asymptotically stable by Theorem 6.5, provided the inequality $(1 + ax^*)^2 > a\beta Ks/d\alpha\gamma$ holds.

The preceding discussion introduces the possibility of stabilization of the interior equilibrium point of a two-dimensional system by the addition of a third trophic level. Suppose, for example, that $\hat{x} < \overline{x} = (aK - 1)/2$. Then for the two-species system the interior equilibrium point $(\hat{x},\hat{y})$ is unstable. The introduction of a third trophic level with an interior equilibrium (E4), (x^*,y^*,z^*), with $x^* > \overline{x}$, introduces an asymptotically stable interior equilibrium into the three-dimensional system.

6.3 ONE PREDATOR AND SEVERAL PREY

In this section the case of one predator feeding on two prey populations will be discussed. Here it is assumed that the two prey populations do not interact with each other (i.e., all competition effects between the prey populations are ignored). This, of course, is only a first approximation to reality and more complicated models could be considered where the prey populations compete (see the Exercises). However, the model about to be discussed is realistic where the two populations (so far as they are concerned) are in isolated habitats.

It will further be assumed that the predator has no prey preference and has free access to the prey, i.e., any isolation of habitats is isolation for the prey only (perhaps the prey crawl and the predators fly). The no preference would also apply in the two-habitat case. For a discussion of models with switching see the Notes on the Literature and Exercises.

The general model then to be considered is

$$\begin{aligned} x_1' &= x_1 g_1(x_1) - y p_1(x_1) \\ x_2' &= x_2 g_2(x_2) - y p_2(x_2) \\ y' &= y[-\gamma + c_1 p_1(x_1) + c_2 p_2(x_2)] \end{aligned} \tag{6.52}$$

where $g_i(x_i)$, and $p_i(x_i)$, $i = 1, 2$, have the usual specific growth rate and predator response functional properties. The third equation of system (6.52) is again a first approximation. For a more general model, again see the Exercises.

The interesting question for the preceding model is that of survival in time of all three species. This boils down to a question of existence and stability of recurrent motions (which include equilibria and periodic motions as special cases) in the interior of the first octant. At this time only equilibria will be considered and it will be seen that, even so, the analysis is nontrivial and only general statements can be made as yet.

For an interior equilibrium to exist, there must exist a positive solution to the simultaneous equation

$$c_1p_1(x_1) + c_2p_2(x_2) = \gamma$$
$$\frac{x_1g_1(x_1)}{p_1(x_1)} = \frac{x_2g_2(x_2)}{p_2(x_2)} \tag{6.53}$$

Positive solutions of (6.53) do not always exist. Indeed, the second equation of (6.53) does not necessarily define a continuous arc in the x_1-x_2 plane under the usual assumptions. However, suppose conditions are such that (6.53) has a solution. Let E: (x_1^*, x_2^*, y^*) be this equilibrium, where

$$y^* = \frac{x_i^* g(x_i^*)}{p_i(x_i^*)}, \qquad i = 1, 2 \tag{6.54}$$

Then the variational matrix at E is given by

$$M = \begin{bmatrix} H_1(x_1^*) & 0 & -p_1(x_1^*) \\ 0 & H_2(x_2^*) & -p_2(x_2^*) \\ c_1y^*p_{1x_1}(x_1^*) & c_2y^*p_{2x_2}(x_2^*) & 0 \end{bmatrix} \tag{6.55}$$

where

$$H_i(x_i^*) = x_i^*g_{ix_i}(x_i^*) + g_i(x_i^*) - \frac{x_i^*g_i(x_i^*)p_{ix_i}(x_i^*)}{p_i(x^*)} \tag{6.56}$$

The characteristic polynomial for M is then

$$\begin{aligned} f(\lambda) = \lambda^3 &- [H_1(x_1^*) + H_2(x_2^*)]\lambda^2 + [H_1(x_1^*)H_2(x_2^*) \\ &+ c_1y^*p_1(x_1^*)p_{1x_1}(x_1^*) + c_2y^*p_2(x_2^*)p_{2x_2}(x_2^*)]\lambda \\ &- [c_1y^*p_1(x_1^*)p_{1x_1}(x_1^*)H_2(x_2^*) \\ &+ c_2y^*p_2(x_2^*)p_{2x_2}(x_2^*)H_1(x_1^*)] \end{aligned} \tag{6.57}$$

By direct substitution it can be seen that

$$f[H_1(x_1^*)]f[H_2(x_2^*)] \le 0$$

and equals zero if and only if $H_1(x_1^*) = H_2(x_2^*)$. Hence there must always be a root of $f(\lambda)$ between $H_1(x_1^*)$ and $H_2(x_2^*)$ that degenerates to $H_i(x_i^*)$ in the case of equality. Hence a necessary condition for the asymptotic stability of E is that one of $H_i(x_i^*)$, $i = 1, 2$, be negative. It is easy to see that the condition that both $H_i(x_i^*) < 0$, $i = 1, 2$, is a sufficient condition for asymptotic stability. For then all coefficients of $f(\lambda)$ are positive, and we already know of one real negative root between $H_1(x_1^*)$ and $H_2(x_2^*)$. Suppose that $H_1(x_1^*) \ge H_2(x_2^*)$. Then if r is this negative real root, it can be written as

$$r = H_1(x_1^*) - a \tag{6.58}$$

where

$$a = \alpha[H_1(x_1^*) - H_2(x_2^*)] \ge 0$$

for some α, $0 < \alpha < 1$. Now dividing $f(\lambda)$ by $(\lambda - r)$, $f(\lambda)$ can be written

$$f(\lambda) = (\lambda - r)q(\lambda) \tag{6.59}$$

where

$$q(\lambda) = \lambda^2 - [H_2(x_2^*) + a]\lambda + \{a[H_2(x_2^*) - H_1(x_1^*)] + a^2 + c_1y^*p_1(x_1^*)p_{1x_1}(x_1^*) + c_2y^*p_2(x_2^*)p_{2x_2}(x_2^*)\} \tag{6.60}$$

But $H_2(x_2^*) + a < 0$ and $a[H_2(x_2^*) - H_1(x_1^*)] + a^2 = a^2(1/\alpha + 1) \ge 0$. Hence all coefficients of $q(\lambda)$ are positive, and being a quadratic $q(\lambda)$ must have roots with negative real parts.

Of course, in the absence of either prey, the condition $H_i(x_i^*) < 0$ is exactly the condition of asymptotic stability, and so the results of this section are not surprising. However, it may be that $H_1(x_1^*) > 0$, $H_2(x_2^*) < 0$, but asymptotic stability holds, in which case it may be that the addition of a second prey population could stabilize (in three dimensions) an unstable (in two dimensions) equilibrium.

NOTES ON THE LITERATURE

The question of food chains has been considered from an analytic point of view by several authors, including Hausrath (1975), Haussmann (1971), Rescigno and Jones (1972), Rosenzweig (1973b), Saunders and Bazin (1975), and, from a network point of view, Yorke and Anderson (1973). The more complicated question of food webs has been discussed by, among others, Barclay and van den Driessche (1975), May (1973c), Haussmann (1971), and Krapivin (1972) (probabilistically).

There have been observations and experiments dealing with a predator feeding on several prey. Some papers dealing with this are Birkland (1974), Maly (1975), Murdoch (1975), Murdoch, Avery and Smyth (1975). Oaten and Murdoch (1975b) have discussed switching of prey by a predator. The question of one predator with two dispersing prey (possibly a genetic dispersal) has been discussed by P. M. Allen (1975), Cramer and May (1972), and Freedman and Waltman (1977a).

EXERCISES

6.1 (Open) Modify system (6.1) so as to increase the length of the chain and analyze the model. This would be of interest even for simple Lotka-Volterra food chains of length 4 or greater.

6.2 For the model (6.52) in the special case of logistic prey growth and Lotka-Volterra predations, derive conditions for the existence of a positive equilibrium.

6.3 (a) (Open) Analyze a model similar to (6.52) incorporating competition between prey.

(b) (Open) Similarly, incorporate switching for the predator; e.g., the third equation could have the form $y' = y[-\gamma + c_1(x_2)p_1(x_1) + c_2(x_1)p_2(x_2)]$ or even $y' = y[-\gamma + p(x_1,x_2)]$ with appropriate assumptions on $p(x_1,x_2)$.

6.4 (Open) Add more prey to model (6.52) and analyze. Also add more predators so as to make the model a food chain as well.

B: APPLICATIONS

For the remainder of this chapter, applications of predator-prey systems to pest control theory and to immunity will be briefly discussed. Both of these topics have recently received much attention.

The basic models in the pest control sections will be of the continuous intermediate type analyzed in Chap. 4. A perturbed model of this form is shown to satisfy the conditions of the Hopf bifurcation theorem, giving the existence of "small-amplitude" periodic solutions.

A not too dissimilar technique is utilized in analyzing a model simulating an immune response. Once the model has been set up and transformed, it is seen to be of the form of a perturbed Kolmogorov type, and again the Hopf bifurcation theorem will give "small-amplitude" periodic solutions.

The sections on the pest control are based on work done by the author; the sections on immunity are based on the work of George Pimbley, Jr.

6.4 PEST CONTROL IN A CONSTANT ENVIRONMENT

In work done in California strawberry patches (Huffaker and Kennett, 1956), in California grapewines (Flaherty and Huffaker, 1970a), and in Tanzania coffee fields (Bigger, 1973), it was observed that for best control of pests by a natural enemy, the pest-enemy populations should be as near equilibrium as possible. Certain experiments were tried in which at the beginning of the growing season, when the pest populations were low, predators were introduced early. The enemies killed off the pests, starved themselves, or went away, and soon thereafter the pests entered the scene again and caused severe crop damage before they were once more controlled by their enemies.

If system (4.2) is taken as a model of the pest-enemy interactions, then from the analysis of this model in Chap. 4, it is easy to see why the preceding occurred. Orbits initiating at the beginning of the season with low x (pest) values will reach a maximum x value close to carrying capacity before x decreases. In fact, the farther away from equilibrium the orbits initiate, the larger the "amplitude" of the oscillations during any one growing season. Hence we conclude that at the beginning of the season, populations should be initiated as close to equilibrium as possible, and for good pest control an enemy should be chosen (if possible) so that its predator response function is such that x^* (the pest value of equilibrium) is sufficiently small.

However, it may be that the equilibrium (x^*,y^*) is unstable; hence one must be concerned that as the orbits initiating near the equilibrium wind away from equilibrium, their amplitudes do not become too large. To investigate this question, system (2.1) will first be modified slightly so that system (4.2) becomes

$$\begin{aligned} x' &- xg(x) - yp(x) \\ y' &= y[-s + cp(x)] \end{aligned} \tag{6.61}$$

We now note from (4.19) that the variational matrix of system (6.61) is

$$\begin{bmatrix} H(x^*) & -p(x^*) \\ cy^*p_x(x^*) & 0 \end{bmatrix} \tag{6.62}$$

where

$$H(x^*) = x^*g_x(x^*) + g(x^*) - \frac{x^*g(x^*)p_x(x^*)}{p(x^*)} \tag{6.63}$$

Now let $p_0(x)$ be such that if (x_0^*,y_0^*) is the corresponding equilibrium of (4.2), then

$$H(x_0^*) = 0 \tag{6.64}$$

and define $\tilde{p}(x)$ by

$$p(x) = p_0(x) + \varepsilon\tilde{p}(x) \tag{6.65}$$

The equilibrium (x^*,y^*) is now a function of ε. Utilizing the implicit function theorem and the fact that $-s + cp_0(x^*) + \varepsilon c\tilde{p}(x^*) = 0$ gives

$$x^*(\varepsilon) = x_0^* - \frac{\tilde{p}(x_0^*)}{p_{0x}(x_0^*)}\varepsilon + o(\varepsilon) \tag{6.66}$$

Then defining $y^*(\varepsilon) = \dfrac{x^*(\varepsilon)g[x^*(\varepsilon)]}{p_0[x^*(\varepsilon)] + \varepsilon\tilde{p}[x^*(\varepsilon)]}$ and expanding in powers of ε gives

$$y^*(\varepsilon) = y_0^* - \frac{y_0^*\tilde{p}(x_0^*)}{p_0(x_0^*)}\varepsilon + 0(\varepsilon) \tag{6.67}$$

In general, as ε passes through zero, $H(x^*)$ will change sign. In the one case, then, (x^*,y^*) will be asymptotically stable; in the other case it will be unstable. It is the unstable case that is, of course, of concern. What shall be shown is that generically, as $H(x^*)$ changes from negative to positive, so that (x^*,y^*) changes from stable to unstable, there will occur "small-amplitude" periodic solutions which are asymptotically stable. The tool utilized is the Hopf bifurcation theorem (see Sec. F of the Appendix).

First we define the quantities

$$\begin{aligned} A = {} & y_0^*\left[\frac{p_{0x}(x_0^*)\tilde{p}(x_0^*)}{p_0(x_0^*)} + \tilde{p}_x(x_0^*)\right] \\ & - \frac{p_0(x_0^*)}{p_{0x}(x_0^*)}\,[x_0^*g_{xx}(x_0^*) + 2g_x(x_0^*) - y_0^*p_{0xx}(x_0^*)] \end{aligned} \tag{6.68}$$

$$B = \frac{d}{dx}\left[\frac{xg_{xx}(x) + 2g_x(x) - y_0^* p_{0xx}(x)}{p_0(x)p_{0x}(x)}\right]_{x=x_0^*} \tag{6.69}$$

Then it will be assumed from now on that

$$A \neq 0, \qquad B < 0 \tag{6.70}$$

Let $M(\varepsilon)$ be the variational matrix M with x^* and y^* as functions of ε given by (6.66) and (6.67). Then

$$M(\varepsilon) = \begin{bmatrix} H[x^*(\varepsilon)] & -p[x^*(\varepsilon)] \\ cy^*(\varepsilon)p_x[x^*(\varepsilon)] & 0 \end{bmatrix} \tag{6.71}$$

First, at $\varepsilon = 0$

$$M(0) = \begin{bmatrix} 0 & -p(x_0^*) \\ cy_0^* p_x(x_0^*) & 0 \end{bmatrix} \tag{6.72}$$

and hence

$$\operatorname{tr} M(0) = 0, \qquad \det M(0) > 0 \tag{6.73}$$

This is the first Hopf hypothesis. For the second hypothesis, $\operatorname{tr} M_\varepsilon(0)$, is needed. But $\operatorname{tr} M_\varepsilon(\varepsilon) = H_\varepsilon[x^*(\varepsilon)]$, and after some computations we get

$$\operatorname{tr} M_\varepsilon(0) = A \neq 0 \tag{6.74}$$

Hence there will appear "small-amplitude" periodic solutions, but it is not yet known whether they will occur when (x^*,y^*) is asymptotically stable (in which case they will be unstable), when $H(x_0^*) = 0$ (in which case they will be neutral), or when (x^*,y^*) is unstable (in which case they will be asymptotically stable). However, in Sec. F of the Appendix a criterion for the stability or instability of these periodic solutions is given. If the parameters of system (6.61) is substituted into this criterion, then after some simplification it is seen that the condition for an asymptotically stable limit cycle of small amplitude when (x^*,y^*) is unstable (and ε sufficiently small) is $B < 0$.

As $|\varepsilon|$ gets larger, it is expected that the amplitude will grow. Hence for best control, if (x^*,y^*) is unstable, $|\varepsilon|$ should not be "too large."

6.5 PEST CONTROL IN AN INCREASINGLY ENRICHED ENVIRONMENT

The analysis of the last section was based on a constant environment. Of course, as the season progresses, one would expect the environment to be more lush. It was shown by Rosenzweig (1971) that this enrichment (modeled by an increase in carrying capacity) could lead to destabilization of a stable equilibrium. He was concerned that this destabilization could lead to possible extinction. It was shown that for our model (which includes Rosenzweig's models) there is always a globally attracting invariant set. However, it is of some concern that destabilization could lead to a "large-amplitude" limit cycle. As in the previous section, it will be shown that such is not the case, for when the carrying capacity increases beyond the critical value causing destabilization, the Hopf bifurcation theorem will once more apply and give the existence of "small-amplitude" stable limit cycles.

We first modify our model so as to acknowledge explicitly the carrying capacity by writing it as

$$\begin{aligned} x' &= xg(x,K) - yp(x) \\ y' &= y[-s + cp(x)] \end{aligned} \tag{6.75}$$

Otherwise all the usual hypotheses on g and p are assumed. We now state the appropriate behaviors of g with respect to K, which generalizes logistic growth, namely, we assume

$$\begin{aligned} &g(0,K) = \alpha > 0, \qquad g(K,K) = 0, \qquad g_x(x,K) \le 0 \\ &g_K(x,K) \ge 0, \qquad g_{xK}(x,K) \ge 0, \qquad x \ge 0 \end{aligned} \tag{6.76}$$

The analysis of Chap. 4 continues to hold, and we see that x^* and $\bar{y}$, as given by (4.32), are independent of K, but y_0, given by (4.30), is now a function of K, and from (4.30)

$$y_0(K) = \frac{-x^*p(x^*)g_x(x^*,K)}{g(x^*,K)} \tag{6.77}$$

We now consider the effect on the model as the carrying capacity increases. We compute $dy_0(K)/dK$.

$$\frac{dy_0(K)}{dK} = \frac{-x^*p(x^*)g_{xK}(x^*,K)}{g(x^*,K)} + \frac{x^*p(x^*)g_x(x^*,K)g_K(x^*,K)}{g(x^*,K)^2} \tag{6.78}$$

Hence by our hypotheses, including (6.76),

$$\frac{dy_0(K)}{dK} \leq 0 \tag{6.79}$$

i.e., as K increases, y_0 decreases, and in fact could move below $\bar{y}$ if it is above $\bar{y}$, which is equivalent to the result obtained by Rosenzweig (1971) that the local maximum could move to the right and possibly destabilize the equilibrium.

It is now supposed that this is the case, i.e., that there exists $K_0 > 0$ such that

$$H(x^*,K_0) = 0 \tag{6.80}$$

where H is given by (6.63), but that

$$g_K(x^*,K_0)^2 + g_{xK}(x^*,K_0)^2 > 0 \tag{6.81}$$

so that H is negative for values of K just to the left of K_0 and positive for values just to the right. We now introduce the parameter ε by

$$\varepsilon = K - K_0$$

We again note that x^* is independent of ε, but that y^* depends on ε, and consider the variational matrix about (x^*,y^*), but now as a function of $K(\varepsilon)$:

$$M(K) = \begin{bmatrix} x^*g_x(x^*,K) + g(x^*,K) - y^*(K)p_x(x^*) & -p(x^*) \\ cy^*(K)p_x(x^*) & 0 \end{bmatrix} \tag{6.82}$$

and we note the following three properties of $M(K)$

1. $\operatorname{tr} M(K_0) = H(x^*, K_0) = 0$
2. $\det M(K_0) = cy^*(K_0)p(x^*)p_x(x^*) > 0$
3. $\operatorname{tr} M_\varepsilon(K_0) = x^* g_{xK}(x^*, K_0) + g_K(x^*, K_0) \neq 0$

Conditions 1, 2, and 3 are again precisely what is required for the Hopf bifurcation theorem to hold as K passes K_0, postulating the existence of "small-amplitude" periodic solutions for either $\varepsilon > 0$, $\varepsilon < 0$, or $\varepsilon = 0$.

We would again like to obtain a criterion that, when satisfied, would guarantee that the small-amplitude periodic solutions occur for $\varepsilon > 0$. The criterion is the same as that used in the previous section, and in fact at $\varepsilon = 0$ the two modified systems are identical. Hence the criterion derived there, which in the modified notation is

$$B = \frac{d}{dx} \left. \frac{x g_{xx}(x, K_0) + 2 g_x(x, K_0) - y^*(K_0) p_{xx}(x)}{c p(x) p_x(x)} \right|_{x=x^*} < 0 \tag{6.83}$$

is again what is needed. If (6.83) holds, it will automatically imply that the small-amplitude periodic solutions are stable. If the opposite inequality holds, the small-amplitude periodic solutions will occur for $\varepsilon < 0$ and will be unstable.

We remark at this time that inequality (6.83) holds in the case of a Lotka-Volterra system with a logistic term introduced, and also for the other systems presented by Rosenzweig (1969, 1971, 1973a) and Rosenzweig and MacArthur (1963).

We now make one last remark, namely, that if destabilization does become "too large", it can be compensated for by the introduction of sources as in Sec. 4.7.

6.6 IMMUNITY: A TWO-DIMENSIONAL MODEL

The final topic of this chapter involves application of predator-prey systems to immune responses. The material in this and the next section is due to George Pimbley, Jr., and is based on a model of G. I. Bell.

Let $u(s)$ be the concentration of antigen and $v(s)$ the concentration of antibody in the system. It is assumed that self-reproducing antigent is introduced into the system, and that antibody, which binds with the antigen, is secreted by lymphocytes as a consequence. Let then $u_b(s)$ and $v_b(s)$ be the bound antigen and antibody, respectively. Then the system modeling the antigen-antibody interaction is given by

$$\begin{aligned} \frac{du}{ds} &= \lambda_1 u - \alpha_1 v_b \\ \frac{dv}{ds} &= -\lambda_2 v + \alpha_2 v_b\left(1 - \frac{v}{\theta}\right) \end{aligned} \tag{6.84}$$

where θ is the carrying capacity for antibody. Using the law of mass action (see Pimbley, 1974a),

$$u_b = v_b = k(u - u_b)(v - v_b) \tag{6.85}$$

and the rational approximation

$$v_b = \frac{kuv}{1 + k(u + v) - kv_b} \approx \frac{kuv}{1 + k(u + v)} \tag{6.86}$$

we arrive at the system

$$\begin{aligned} \frac{du}{ds} &= u\left[\frac{\lambda_1 + k\lambda_1 u - k(\alpha_1 - \lambda_1)v}{1 + k(u + v)}\right] \\ \frac{dv}{ds} &= v\left[\frac{-\lambda_2 + k(\alpha_2 - \lambda_2)u - k\lambda_2 v - (k\alpha_2/\theta)uv}{1 + k(u + v)}\right] \end{aligned} \tag{6.87}$$

At this point the change of independent variable

$$t = \int_0^s [1 + k(u(s') + v(s'))]^{-1}\, ds' \tag{6.88}$$

is made. This modifies system (6.87) to the form

$$
\begin{aligned}
x' &= x[\lambda_1 + k\lambda_1 x - k(\alpha_1 - \lambda_1)y] \\
y' &= y\left[-\lambda_2 + k(\alpha_2 - \lambda_2)x - k\lambda_2 y - \frac{k\alpha_2}{\theta}\, xy\right]
\end{aligned}
\tag{6.89}
$$

where $' = d/dt$, $\alpha_1 > 0$, $\alpha_2 > 0$, $\lambda_1 > 0$, $\lambda_2 > 0$, $\theta > 0$.

Equilibria are at the intersections of the isoclines of system (6.89). In actual fact, these isoclines may not intersect. However, sufficiency conditions which guarantee such equilibria are

$$
\begin{aligned}
&\alpha_1 > \lambda_1 + \lambda_2, \qquad \alpha_2 > \lambda_1 + \lambda_2 \\
&\theta > \frac{\lambda_1/k}{(\sqrt{\alpha_m - \lambda_1} - \sqrt{\lambda_2})^2}
\end{aligned}
\tag{6.90}
$$

where $\alpha_m = \min(\alpha_1, \alpha_2)$. If (6.90) holds, in general there will be two interior equilibria, one of which is a focus, center, or node, and the other of which is a saddle point. We will let (x^*, y^*) denote the first of these equilibria.

At this time we will say what we mean by an immune response. We mean the existence of "small-amplitude" asymptotically stable limit cycles. This would imply the continued existence of antibody interacting with the antigen, so that if there were a change of initial conditions (by the introduction of more antigen), the effect of the antibody would be to reduce the antigen and move the interaction trajectory toward the limit cycle.

The technique used is Waltman's as described in Sec. 5.4. Instead of (6.89) we consider

$$
\begin{aligned}
x' &= x[\lambda_1 + k\lambda_1 x - k(\alpha_1 - \lambda_1)y] \\
y' &= \beta y\left[-\lambda_2 + k(\alpha_2 - \lambda_2)x - k\lambda_2 y - \frac{k\alpha_2}{\theta}\, xy\right]
\end{aligned}
\tag{6.91}
$$

Then if $\beta = \beta_0 + \varepsilon$, where

$$
\beta_0 = \frac{\lambda_1 x^*}{\lambda_2 y^* + (\alpha_2/\theta)x^* y^*}
$$

the conditions of the Hopf bifurcation theorem are again seen to hold at $\varepsilon = 0$. Further, it is seen that the "small-amplitude" periodic solutions occur for $\varepsilon < 0$ and are asymptotically stable, provided

$$\beta_0 > \frac{\lambda_1}{\alpha_2 - \lambda_2} \tag{6.92}$$

Further, there exists a value α_{10} of α_1 such that as α_1 is increased through α_{10}, with α_2, λ_1, λ_2, and θ fixed, β_0 is increased through unity. For details see Pimbley (1974a).

6.7 IMMUNITY: A THREE-DIMENSIONAL MODEL

In this section it is assumed that in addition to the presence of antigen and antibody in the immune response, there are also lymphocytes which serve a dual purpose. First, it is the lymphocytes that secrete the antibody, and second, the lymphocytes have receptors capable of binding to the antigen. It is assumed that each lymphocyte has n receptors.

After a similar set of assumptions and transformations as in the previous section, the system of equations to be considered is

$$\begin{aligned}
x' &= x[\lambda_1 + k\lambda_1 x - k(\alpha_1 - \lambda_1)y + kn\lambda_1 z] \\
y' &= \beta\{y[-\lambda_2 - k(\alpha_2 + \lambda_2)x - k\lambda_2 y - kn\lambda_2 z] + k\gamma xz\} \\
z' &= z\left[-\lambda_3 + k(\alpha_3 - \lambda_3)x - k\lambda_3 y - kn\lambda_3 z - \frac{k\alpha_3}{\theta} xz\right] \\
&\quad + S[1 + kx + ky + nkz]
\end{aligned} \tag{6.93}$$

(see Pimbley, 1974c, for the derivation). S is a source term in the creation of lymphocytes.

The techniques used to show the existence of "small-amplitude" periodic solutions are precisely the same as in the previous section for $S = 0$. First it is shown that there is a $\beta_0 > 0$ such that at $\beta = \beta_0$, the hypotheses of the Hopf bifurcation theorem hold.

When $\lambda_3 = 0$, the direction of bifurcation is also obtained. Then it is shown that by varying one other parameter, say α_1, there is a critical value at which $\beta_0 = 1$.

Hsu and Kazarinoff (1977) have obtained results on the direction of bifurcation when $\lambda_3 > 0$ and have considered (6.93) when $S > 0$.

NOTES ON THE LITERATURE

Papers dealing with mathematical models and analysis of pest control go back to Nicholson (1933). Also see Bigger (1973, 1976), Diamond (1974a), Hassell and Varley (1967), Nicholson and Bailey (1935), Rudd (1975), and Watt (1959, 1961). For papers dealing with the enrichment aspect of predator-prey systems see Gilpin (1972), McAllister et al. (1972), and Rosenzweig (1971, 1972a,b).

There are many works dealing with actual attempts at pest control by a natural enemy as well as observations of the effects of insecticides. Some of these are Beirne (1975), Ehler and van den Bosch (1974), Flaherty (1969), Flaherty and Huffaker (1970a,b), Ghabbour (1972-74), Huffaker and Kennett (1956), McMurtry and van de Vrie (1973), Polyakov (1972), Wallace and Walters (1974).

The mathematical treatment of immunity began with the work of Bell (1970, 1971a,b, 1973, 1974). Based on Bell's models, Pimbley (1974a,b,c) and Hsu and Kazarinoff (1977) have analyzed models with replicating antigen. Waltman and Butz (1977) and Gatica and Waltman (1976) have looked at threshold models for B-lymphocyte response to specific nonreplicating antigen. Perelson et al. (1976) consider immune response from an optimal strategy point of view, whereas Richter (1975) has looked at immunity from a network point of view. Other aspects of immunity may be found in, e.g., Bruni et al. (1975), S. Cohen (1971), Jaroszewski et al. (1976), Jerne (1973, 1974), and Raff (1976). See Merrill (1976a) for a nice summary.

EXERCISES

6B.1 (Open) In system (6.61) incorporate time lags in the predator response function and perturbations and determine how they affect the stability of the system.

6B.2 (Open) In system (6.84) incorporate periodic source terms in the model and analyze. Do the same for system (6.93).

PART III
COMPETITION AND COOPERATION (SYMBIOSIS)

In the final three chapters of this book the topics of competition and cooperation will be considered. In Chap. 7, Lotka-Volterra competition models will be analyzed, and the competitive exclusion principle discussed. In Chap. 8, when higher-degree models are considered, it will be shown that exclusion will not necessarily be the case.

Kolmogorov-type models for both competition and cooperation will be analyzed. In addition, there will be a short section on niche theory.

Chapter 7
LOTKA–VOLTERRA COMPETITION MODELS

7.1 INTRODUCTION

As in predator-prey models, the simplest models of competition are of the Lotka-Volterra type. It is assumed as always that the growth of each species is proportional to the number of that species present. The "proportionality constant" has a positive contribution from the presence of that species, at least up to carrying capacity (if such exists), and a nonpositive (usually negative) contribution from the presence of one or more other species.

When we say that the growth model is in general of the Lotka-Volterra type, we mean that the growth equation of the species x_i is of the form

$$x_i' = x_i f_i(x_1, \ldots, x_n), \qquad i = 1, \ldots, n$$

where f_i is a linear function of its variables.

In the next several sections we will consider the case when $n = 2$ and then give some discussion for $n > 2$. We finish the chapter with a short discussion of niche theory.

7.2 THE TWO-DIMENSIONAL MODEL WITHOUT CARRYING CAPACITY

In this section we examine the two-dimensional Lotka-Volterra competition model without carrying capacity. This model has the form

$$\begin{aligned} x_1' &= x_1(\alpha_1 - \beta_1 x_2) \\ x_2' &= x_2(\alpha_2 - \beta_2 x_1) \end{aligned} \tag{7.1}$$

where x_1, x_2 are the numbers or densities or biomass of the competing species, α_i is the specific growth rate of x_i, and β_i's are the competition coefficients.

There are several equilibria for this system. Clearly (E1): (0,0) and (E2): $(\alpha_2/\beta_2, \alpha_1/\beta_1)$ are these equilibria. Since each species grows in the absence of the other, (E1) is completely unstable. To analyze (E2), we consider the variational matrix of (7.1) at (E2). Let V_2 be this matrix. Then

$$V_2 = \begin{bmatrix} 0 & -\dfrac{\alpha_2\beta_1}{\beta_2} \\ -\dfrac{\alpha_1\beta_2}{\beta_1} & 0 \end{bmatrix} \tag{7.2}$$

and the eigenvalues of V_2 are

$$\lambda = \pm \sqrt{\alpha_1\alpha_2} \tag{7.3}$$

Hence (E2) is a hyperbolic point.

System (7.1) can be solved in the phase plane. First one gets the equation

$$\frac{dx_2}{dx_1} = \frac{x_2(\alpha_2 - \beta_2 x_1)}{x_1(\alpha_1 - \beta_1 x_2)} \tag{7.4}$$

or $(\alpha_1 x_2^{-1} - \beta_1)\, dx_2 = (\alpha_2 x_1^{-1} - \beta_2)\, dx_1$. Letting x_{10} and x_{20} be the initial populations at $t = 0$, integration gives

$$\alpha_1 \ln \left[\frac{x_2}{x_{20}}\right] - \beta_1(x_2 - x_{20}) = \alpha_2 \ln \left[\frac{x_1}{x_{10}}\right] - \beta_2(x_1 - x_{10}) \tag{7.5}$$

From Eq. (7.5) it is seen that except for curves initiating on a certain separatrix Γ (see Fig. 2.1), either x_1 or x_2 approaches zero as $t \to \infty$. To obtain the equation of Γ, we note that the left member reaches its maximum at $x_2 = \alpha_1/\beta_1$, and the right member reaches its maximum when $x_1 = \alpha_2/\beta_2$. The separatrix occurs when the initial values have the critical values just mentioned, i.e., Γ is given by

$$\alpha_1 \ \ell n \left[\frac{\beta_1 x_2}{\alpha_1}\right] - \beta_1 x_2 + \alpha_1 = \alpha_2 \ \ell n \left[\frac{\beta_2 x_1}{\alpha_2}\right] - \beta_2 x_1 + \alpha_2 \tag{7.6}$$

The population dynamics are illustrated in Fig. 7.1.

7.3 THE TWO-DIMENSIONAL MODEL WITH ONE CARRYING CAPACITY

We consider in this section the system

$$\begin{aligned} x_1' &= x_1\left(\alpha_1 - \frac{\alpha_1 x_1}{K_1}\right) - \beta_1 x_1 \\ x_2' &= x_2(\alpha_2 - \beta_2 x_1) \end{aligned} \tag{7.7}$$

Here the species x_1 has a carrying capacity associated with it, K_1.

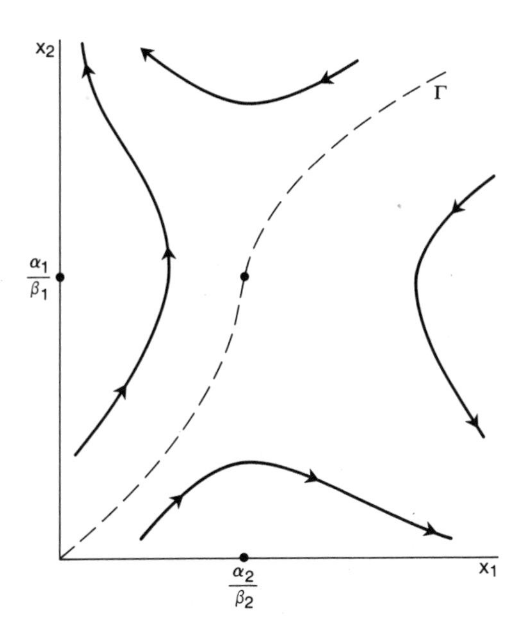

Fig. 7.1 The separatrix Γ in the case of no carrying capacities. Here $\alpha_2/\beta_2 > \alpha_1/\beta_1$. In the opposite case, the concavities are reversed.

We note that there are always three equilibria for system (7.7), namely, (E1): (0,0), (E2): $(K_1,0)$, and (E3): $(\alpha_2/\beta_2, \alpha_1/\beta_1\ [1-(\alpha_2/\beta_2K_1)])$. (E3) is positive if and only if $\alpha_2/\beta_2 < K_1$.

Let M(x,y) be the variational matrix of system (7.7) and M_i be this matrix evaluated at (Ei). Then

$$M(x,y) = \begin{bmatrix} \alpha_1 - \dfrac{2\alpha_1 x_1}{K_1} - \beta_1 x_2 & -\beta_1 x_1 \\ \\ -\beta_2 x_2 & \alpha_2 - \beta_2 x_1 \end{bmatrix} \tag{7.8}$$

and

$$M_1 = \begin{bmatrix} \alpha_1 & 0 \\ \\ 0 & \alpha_2 \end{bmatrix}, \quad M_2 = \begin{bmatrix} -\alpha_1 & -\beta_1 K_1 \\ \\ 0 & \alpha_2 - \beta_2 K_1 \end{bmatrix}$$

$$M_3 = \begin{bmatrix} \dfrac{-\alpha_1\alpha_2}{\beta_2 K_1} & \dfrac{-\alpha_2\beta_1}{\beta_2} \\ \\ -\dfrac{\alpha_1\beta_2}{\beta_1}\left(1 - \dfrac{\alpha_2}{\beta_2 K_1}\right) & 0 \end{bmatrix} \tag{7.9}$$

The stability analysis follows readily from (7.9). Clearly the origin is unstable. In the absence of either species, the other species grows. For the equilibrium (E2), the x_1 direction is always asymptotically stable. The x_2 direction is stable or unstable corresponding to $\alpha_2 - \beta_2K_1$ being negative or positive, respectively.

For (E3) we are interested in the case where $\alpha_2/\beta_2 < K_1$. In this case it is easy to see that with the eigenvalues of M_3, both are real, one positive and one negative. Hence in this case (E3) is a hyperbolic point.

We can actually get a picture of the global dynamics for system (7.7). If $x_1 < \alpha_2/\beta_2$, then $x_2(t)$ increases, whereas if $x_1 > \alpha_2/\beta_2$, then $x_2(t)$ decreases. Further, if x_1 and x_2 are above the line

$$\frac{\alpha_1}{K_1} x_1 + \beta_1 x_2 = \alpha_1 \tag{7.10}$$

then $x_1(t)$ decreases, whereas if they are below (7.10), then $x_1(t)$ increases.

From the preceding, it is clear that if $\alpha_2/\beta_2 \geq K_1$, then $x_1(t)$ is always driven extinct. If $\alpha_2/\beta_2 < K_1$, there again exists a separatrix Γ passing through the origin and through (E3), such that

1. For populations initiating to the left of Γ, $x_1(t) \to 0$, $x_2(t) \to \infty$.
2. For populations initiating to the right of Γ, $x_1(t) \to K_1$, $x_2(t) \to 0$.
3. For populations initiating on Γ, $[x_1(t), x_2(t)] \to$ (E3).

The dynamics for the various cases are illustrated in Fig. 7.2.

If the system is of the form

$$x_1' = x_1(\alpha_1 - \beta_1 x_2)$$

$$x_2' = x_2\left(\alpha_2 - \beta_2 x_1 - \frac{\alpha_2 x_2}{K_2}\right)$$

the roles of x_1 and x_2 in the preceding analysis are interchanged.

7.4 THE TWO-DIMENSIONAL MODEL WITH TWO CARRYING CAPACITIES

Here it is supposed that both competing species have carrying capacities. The model becomes

$$\begin{aligned} x_1' &= x_1\left(\alpha_1 - \frac{\alpha_1 x_1}{K_1} - \beta_1 x_2\right) \\ x_2' &= x_2\left(\alpha_2 - \beta_2 x_1 - \frac{\alpha_2 x_2}{K_2}\right) \end{aligned} \tag{7.11}$$

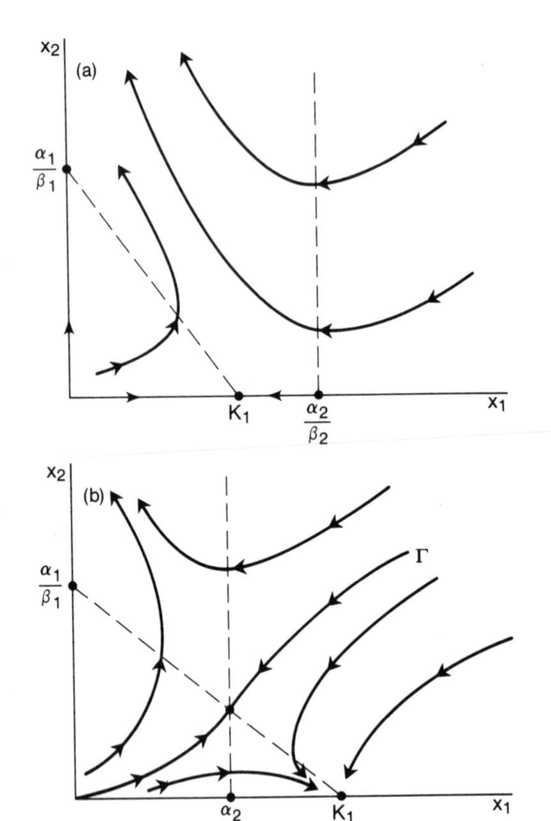

Fig. 7.2 (a) The dynamics of system (7.7) when $\alpha_2/\beta_2 > K_1$. x_1 always goes extinct. (b) $\alpha_2/\beta_2 < K_1$. (E3) is positive. Only for solutions initiating on Γ does extinction not occur.

There are now four equilibria, (E1): (0,0), (E2): $(K_1,0)$, (E3): $(0,K_2)$, and (E4): $[\alpha_2K_1(\alpha_1-\beta_1K_2)/(\alpha_1\alpha_2-\beta_1\beta_2K_1K_2), \alpha_1K_2(\alpha_2-\beta_2K_1)/(\alpha_1\alpha_2-\beta_1\beta_2K_1K_2)]$. (E4), of course, exists if and only if $\alpha_1\alpha_2 - \beta_1\beta_2K_1K_2 \neq 0$. For the sake of notation let

$$\Delta = \alpha_1\alpha_2 - \beta_1\beta_2K_1K_2 \tag{7.12}$$

The variation matrix of (7.11) is

$$M(x,y) = \begin{bmatrix} \alpha_1 - \dfrac{2\alpha_1 x_1}{K_1} - \beta_1 x_2 & -\beta_1 x_1 \\ -\beta_2 x_2 & \alpha_2 - \beta_2 x_1 - \dfrac{2\alpha_2 x_2}{K_2} \end{bmatrix} \tag{7.13}$$

Letting M_i be the value of $M(x,y)$ at (Ei), we get

$$M_1 = \begin{bmatrix} \alpha_1 & 0 \\ 0 & \alpha_2 \end{bmatrix}$$

$$M_2 = \begin{bmatrix} -\alpha_1 & -\beta_1 K_1 \\ 0 & \alpha_2 - \beta_2 K_1 \end{bmatrix}$$

$$M_3 = \begin{bmatrix} \alpha_1 - \beta_1 K_2 & 0 \\ -\beta_2 K_2 & -\alpha_2 \end{bmatrix}$$

$$M_4 = \begin{bmatrix} \dfrac{-\alpha_1\alpha_2(\alpha_1 - \beta_1 K_2)}{\Delta} & \dfrac{-\alpha_2\beta_1 K_1(\alpha_1 - \beta_1 K_2)}{\Delta} \\ \dfrac{-\alpha_1\beta_2 K_2(\alpha_2 - \beta_2 K_1)}{\Delta} & \dfrac{-\alpha_1\alpha_2(\alpha_2 - \beta_2 K_1)}{\Delta} \end{bmatrix} \tag{7.14}$$

We consider now various cases, noting that on the axes, both species approach their respective carrying capacities.

CASE A. $\alpha_1/\beta_1 < K_2$, $\alpha_2/\beta_2 < K_1$. In this case $\Delta < 0$ and (E4) is in the first quadrant. Here M_2 and M_3 attract in the positive directions, and in fact there is once more a separatrix Γ such that everything on one side of it approaches K_1 and on the other approaches K_2. (E4) is an unstable node (see Fig. 7.3a).

CASE B. $\alpha_1/\beta_1 > K_2$, $\alpha_2/\beta_2 > K_1$. In this case (E2) and (E3) repel in the x_2 and x_1 directions, respectively, whereas (E4) is a stable attractor in the first quadrant ($\Delta > 0$). All solutions initiating in the first quadrant approach (E4) (see Fig. 7.3b).

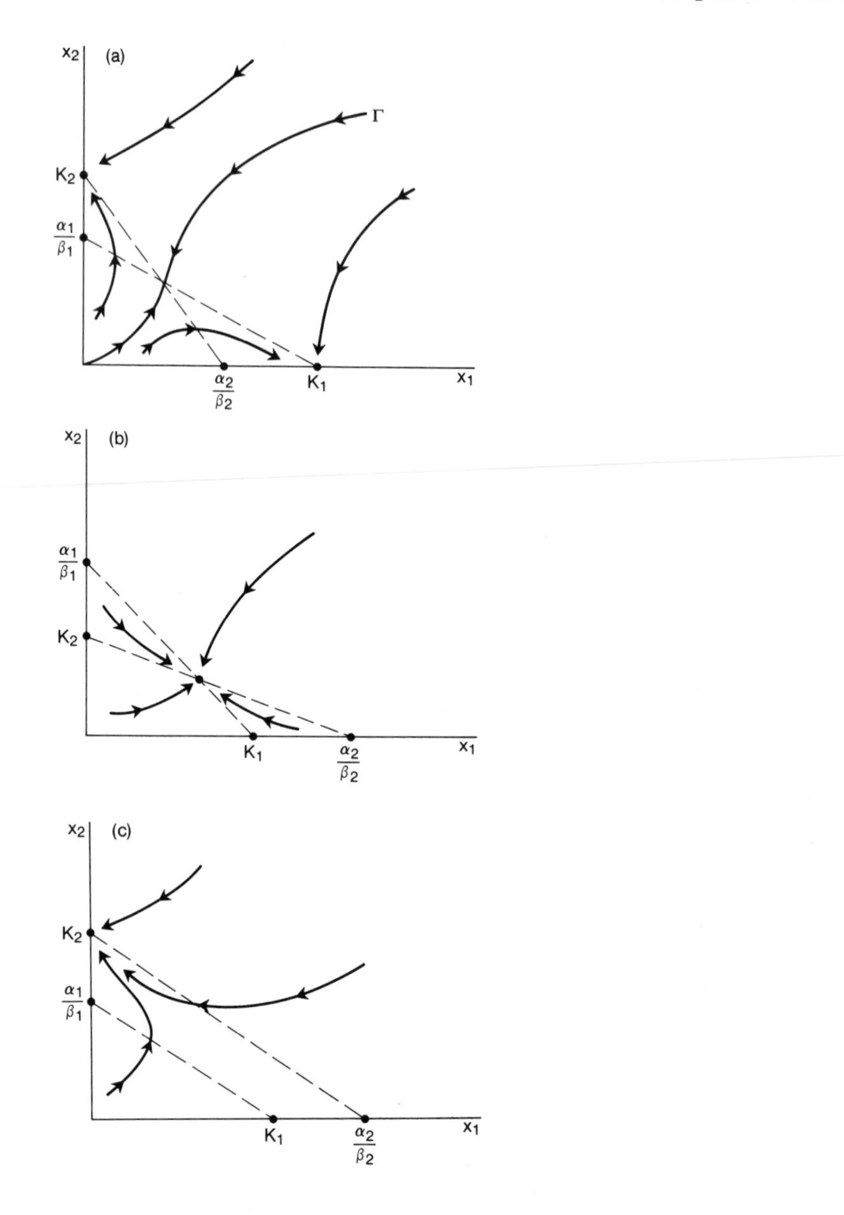

Fig. 7.3 (a) The dynamics of system (7.11) case A. All solutions to the right of Γ tend to (E2) whereas all solutions to the left of Γ tend to (E3). (b) The dynamics of system (7.11), case B. All solutions tend to (E4). (c) The dynamics of system (7.11), case C. All solutions tend to (E3).

CASE C. $\alpha_1/\beta_1 < K_2$, $\alpha_2/\beta_2 \geq K_1$. In this case (E2) repels in the x_2 direction, (E3) attracts in the x_1 direction, and (E4), if it exists, is not in the interior of the first quadrant. In this case all solutions initiating in the interior of the first quadrant tend to (E3) (see Fig. 7.3c).

CASE D. $\alpha_1/\beta_1 \geq K_2$, $\alpha_2/\beta_2 < K_1$. This case is similar to the previous one, with all solutions tending to (E2)

7.5 HIGHER-DIMENSIONAL MODELS: THE COMPETITIVE EXCLUSION PRINCIPLE

In this section we consider the case where there are n species in competition for r resources. The ideas in this section are due to S. A. Levin and may be found in Levin (1970).

Suppose there are n species, $x_1, \ldots, x_n$ and m parameters, $u_1, \ldots, u_m$, representing environmental factors affecting species growth but unaffected by the species dynamics. Let the dynamics of species growth be given by

$$x_i' = x_i f_i(x_i, \ldots, x_n, u_1, \ldots, u_m), \qquad (7.15)$$
$$i = 1, \ldots, n$$

Other than sufficient smoothness, no assumptions are made on the dependence of the f_i or the x_j's and u_j's.

Suppose there is a set of necessary resources $\{z_1, \ldots, z_r\}$, crucial to the survival of the x_i's, and suppose further that each z_i, $i = 1, \ldots, r$, $1 \leq r \leq m + n$, is a function of the x_j's and u_j's, i.e., $z_i = z_i\ (x_1, \ldots, x_n, u_1, \ldots, u_m)$, in such a way that the f_i's can be thought of as functions of the z_j's. Then (7.15) can be written as

$$x_i' = x_i f_i(z_1, \ldots, z_n), \quad i = 1, \ldots, n \qquad (7.16)$$

At this time we make the Lotka-Volterra assumption, namely, that all combinations are such that the f_i's are linear functions of the z_j's, i.e.,

$$f_i(z_1, \ldots, z_r) = \sum_{j=1}^{r} \alpha_{ij} z_j + \beta_i, \quad i = 1, \ldots, n \qquad (7.17)$$

We now consider the special case where $r < n$. Since the f_i's are linear in the z_j's and since there are more functions than unknowns, there is a set of real numbers $\{\gamma_i\}_1^n$, not all zero, such that

$$\sum_{i=1}^{n} \gamma_i f_i(z_1, \ldots, z_r) = c \tag{7.18}$$

where c is a constant. If $r < n - 1$, c can be made equal to zero. Substituting into (7.16) gives

$$\sum_{i=1}^{n} \frac{\gamma_i x_i'}{x_i} = c \tag{7.19}$$

Integrating (7.19) gives

$$\sum_{i=1}^{n} \gamma_i \ln x_i = ct + k \tag{7.20}$$

where k is a constant uniquely determined by the orbit of the solutions. Finally (7.20) can be put into the form

$$x_1^{\gamma_1} \cdots x_n^{\gamma_n} = e^k e^{ct} \tag{7.21}$$

If $c = 0$, the right-hand side of (7.21) is constant and unique for each orbit of solutions. Eq. (7.21) shows that it is impossible for any orbit to approach any other orbit which lies interior to the positive hyperoctant. In particular, this implies that no equilibrium or periodic solution can be asymptotically stable (but they could be stable). In particular, therefore, the ω-limit set of a nonrecurrent motion must, if not unbounded, lie on one of the coordinate hyperplanes.

If $c \neq 0$, then at least one of the x_i's must approach either zero or become unbounded, which again means that no bounded solution can be asymptotically stable.

The preceding is a justification (in the Lotka-Volterra case) of the competitive exclusion principle, which states that "no ecological community in which n species are limited by less than n resources can persist indefinitely."

For a historical introduction to this principle see Levin (1970). We will see in the next chapter that for general systems, this principle does not necessarily hold.

7.6 NICHE THEORY

The ideas and concepts of niche theory properly belong in a book dealing with the stochastic aspects of population ecology, rather than the deterministic, since almost all the work done on niche theory is statistical. However, because of the close relationship between the concept of a niche and the competitive exclusion principle, a short section on niche theory will be presented at this time.

The competitive exclusion principle may be restated as "n species cannot occupy fewer than n niches at any given time." But what precisely is a niche? It is our belief that even today a uniform definition of "niche" has not been attained.

In 1904 Grinell (1904) probably started niche theory (although he did not use the word) when he claimed that intracompetition among bird species existing on common ground forced the species to seek different foods.

According to Gaffney (1975), the word "niche" was first used in an ecological context in Johnson (1910). The first connection between natural selection, competition, and the niche is apparently given in Gause (1934).

The quatification of the niche began with Hutchinson (1957), who viewed the niche as a hypervolume in Euclidean space with axes along each resource utilized by the species. This led to the idea of a "niche width."

Many techniques have been described to measure niche widths and niche overlays, all of them, of course, statistical. One technique (Colwell and Futuyma, 1971) is to consider the niche breadth as being measured by the uniformity of the distribution of individuals among the resource states. Levins (1968) suggested several such measures of uniformity B_i; the niche breadth of the i-th species is given by

$$B_i = \frac{1}{\sum_j p_{ij}^2} \tag{7.22}$$

or

$$B_i = -\sum_j p_{ij} \log p_{ij}$$

where $p_{ij} = x_{ij}/\Sigma_j x_{ij}$, the proportion of the individuals of species i associated with resource state j. Colwell and Futuyma (1971) have suggested incorporating weighting factors in the measurements to account for heterogeneity in the environment.

For a historical introduction to niche theory, see Cody (1974).

NOTES ON THE LITERATURE

There have been many papers written on Lotka-Volterra or near Lotka-Volterra competition models. Among others see Abdelkader (1974), Gomatam (1974b), Ito (1971), Kerner (1961) (three species), Koch (1974a), Leung (1976), MacArthur and Levins (1967), Pielou (1974), Ross (1973a,b), Slatkin (1974), Strobeck (1973), Turner and Rapport (1974), Vandemeer (1970, (for n species), 1975).

Several papers dealing with competitive systems involving time delays are Bartlett (1957), Cushing (1976a), and Gomatam and MacDonald (1975). Gilpin (1974d) has discussed Liapunov functions for competition communities, Yodzis (1976) has considered competition systems with harvesting, and Abrams (1975) has considered methods of computing competition coefficients.

Several papers have been concerned with periodic solutions of near Lotka-Volterra systems. Among others are Gilpin (1975a), Grasman (preprint), and May and Leonard (1975).

Some authors have also considered the question of species packing and competitive systems. Some papers on the subject are MacArthur (1968) and Roughgarden (1974b, 1975b, 1976). In addition, the paper of Levins and Culver (1971) deals with extinction and migration. Riebesell (1974) looks at the enrichment paradox for competitive systems.

There have been very many papers dealing with experiments, observations, simulations, or combinations of these on competitive systems in an attempt to justify or deny the competitive exclusion principle, to determine dominance, or to fit data to theory. Some of these are Beaver (1974), Bovbjerg (1970), Colwell (1973), Culver (1970), De Benedictis (1974), Gilpin (1974c), Gilpin and Ayala (1976), Goudriaan and de Wit (1973), Grenney et al. (1973), Jaeger (1972, 1974), Leslie et al. (1968), M'Closkey (1976), Mertz et al. (1976), Miller (1964, 1968), Petersen (1975), Rathcke (1976), Sheppard (1971), Titman (1976), Vandemeer (1969), Vaughan and Hansen (1974), and Zaret (1972).

At this time we mention that J. Cohen (1970), Roughgarden (1975a), and Schoener (1974b) are papers dealing with statistical aspects of competition.

Some papers dealing with some of the more general aspects of competition are Birch (1957), Hairston et al. (1960), Hoern and MacArthur (1972), Hubbell (1973b), Huffaker and Laing (1972), Hutchinson (1961), Tinnin (1972), and Waldon (1975).

As to niche theory, the number of papers is large. Papers dealing with various aspects of the subject are Abrams (1976), Cody (1974), Colwell and Futuyma (1971), Darlington (1972), Gliddon and Strobeck (1975), Green (1974), Levandowsky (1972), Maguire (1967, 1973), May (1974a), May and MacArthur (1972), McNaughton and Wolf (1970), Milstead (1972), Pielou (1972), Rejmanek and Jenik (1975), Ricklefs (1972), Roughgarden (1974a,c,d), Sabath (1974), Sabath and Jones (1973), Slatken and Lande (1976), Sykes (1974), Thoday (1974), Whittaker et al. (1973), and Zaret and Rand (1971).

EXERCISES

7.1 (Cushing 1976d, 1977a) In systems (7.1), (7.7), and (7.11), let some or all of the coefficients be periodic functions of t. Derive conditions for extinction, coexistence, and periodic solutions.

7.2 (Open) For each of the systems (7.1), (7.7), and (7.11), add perturbational terms of the form $\varepsilon f_i(x_1,x_2)$. Analyze the behavior of solutions in the critical cases.

7.3 Consider the n-dimensional model given by $x_i' = x_i(-\Sigma_{j=1}^n \alpha_{ij}x_j + \beta_i)$, $i, j = 1, 2, \ldots, n$. Determine the equilibria and their stability in this system. Determine conditions for invasion or extinction. (Invasion occurs when small initial populations of one species grow near established stable equilibria of the rest.)

Chapter 8
HIGHER-ORDER COMPETITION MODELS

8.1 INTRODUCTION

In this chapter we are concerned with competition models in which the specific growth rates are nonlinear, as distinct from Lotka-Volterra models discussed in the last chapter, where the specific growth rates are linear. In particular, we will be looking at competing species living on both fixed and renewable resources. Incorporated with this will be an analysis of a system of two predators and one prey.

The competitive exclusion principle will once more be examined and an example will be given to show that indeed this principle may be violated (at least methematically).

The aforementioned analysis and example is due to McGehee and Armstrong (1977).

Finally we look at Komogorov-type analyses and present results due to Rescigno and Richardson (1965), Albrecht et al. (1974), and Rescigno (1968).

8.2 COMPETITION FOR FIXED RESOURCES

In this section we consider the case where the dynamics of the ecosystem is given by n competing species for r fixed resources. By "fixed" we mean that the resources are at a certain fixed value in the absence of species utilizing them. Throughout the next several sections, we will let y_i represent the number (or density or biomass) of the i-th competing species.

At this time we will define what is meant by an attractor block. In general, we will be looking at solutions of a system of differential equations. Without getting into the details of topological dynamics, a closed domain with nonempty interior B in the n-dimensional Euclidean space is an *attractor block* if for all solutions initiating at some time on the boundary of B, for all greater times these solutions will be in the interior of B. Further, a closed set A is called an *attractor* if the ω-limit set of all solutions initiating sufficiently close to A is contained in A. Clearly, every attractor is contained within an attracting block. Similarly, a *repeller block* and a *repeller* are an attractor block and attractor, respectively, for reverse time (i.e., as $t \to -\infty$).

We now suppose that the dynamics of the competing species is given by the system

$$\begin{aligned} y_i' &= y_i f_i(z_i, \ldots, z_r), \qquad i = 1, \ldots, n \\ z_j &= u_j(y_1, \ldots, y_n), \qquad j = 1, \ldots, r \end{aligned} \tag{8.1}$$

where the z_j's are the fixed resources. We will in general say that system (8.1) is *persistent* if there is an attracting block in the interior of the positive hyperoctant in R^n and that it *exhibits the exclusion principle* if it is not persistent.

At this point we shall try to clarify two types of problems for consideration. The first is to consider a specific system of type (8.1) and decide whether it is persistent or not. The second is to consider the set of all allowable systems (8.1) for fixed n and r and decide generically whether this set is persistent or not. McGehee and Armstrong (1977) have been mainly concerned with the second of these types.

It has been shown that if $r \geq n$, persistence generically occurs. However, if $r \leq n$, then almost no system (8.1) has an interior equilibrium, and if it does, that equilibrium is not an attractor. Finally, for the special case $n = 2$, $r = 1$, it can be shown that generically the exclusion property holds. For proofs of the preceding statements see McGehee and Armstrong (1977).

There is as yet not much work done on problems of the first type. Most of the work thus far has been to construct specific examples to illustrate persistence. Nitechi (preprint) has recently shown how to construct an example of system (8.1) for $r = 1$, $n \geq 3$ arbitrary, possessing an attracting periodic orbit.

8.3 COMPETITION FOR RENEWABLE RESOURCES

In this section we consider systems of the form

$$\begin{aligned} y_i' &= y_i f_i(x_1, \ldots, x_r), \qquad i = 1, \ldots, n \\ x_j' &= x_j u_j(y_1, \ldots, y_n, x_1, \ldots, x_r) \qquad j = 1, \ldots, r \end{aligned} \tag{8.2}$$

System (8.2) may be thought of as a system of n predator species competing for r prey species.

The same kinds of questions and definitions raised in the preceding section are valid for system (8.2) as well. In particular, for a given system of type (8.2), the question of whether or not there is an attracting block is an important and open question.

We will be particularly interested in f_i's and u_j's, which are like the intermediate models in Chap. 4 for the purpose of constructing a counterexample to the competitive exclusion principle in the next section.

Before leaving this section we mention the work of S. B. Hsu (1976), who has obtained conditions for exclusion for certain types of competing species of microorganisms that compete for a single nutrient in a chemostat in a manner simulating system (8.2) (see the Exercises).

8.4 AN EXAMPLE OF McGEHEE AND ARMSTRONG: VIOLATION OF THE EXCLUSION PRINCIPLE

We consider the system of two predators and one prey given by the intermediate model

$$\begin{aligned} x' &= xg(x) - y_1p_1(x) - y_2p_2(x) \\ y_1' &= y_1[-m_1 + c_1p_1(x)] \\ y_2' &= y_2[-m_2 + c_2p_2(x)] \end{aligned} \tag{8.3}$$

where $g(x)$, $p_1(x)$, and $p_2(x)$ have the usual properties of predator response as given in Chap. 4 and all constants are positive. Further, we note that system (8.3) has an interior equilibrium only if the system

$$\begin{aligned} -m_1 + c_1p_1(x) &= 0 \\ -m_2 + c_2p_2(x) &= 0 \end{aligned} \tag{8.4}$$

has a solution, an extremely unlikely occurrence. If, however, $x = x^*$ does solve (8.4), there will be a line of equilibria given by

$$p_1(x^*)y_1 + p_2(x^*)y_2 = x^*g(x^*) \tag{8.5}$$

The example of McGehee and Armstrong showing persistence for a system of type (8.3) will be constructed in six stages.

First consider

$$\begin{aligned} x' &= xg(x) - \beta_0xy_1 - \beta_0xy_2 \\ y_1' &= y_1(-m + c\beta_0x) \\ y_2' &= y_2(-m + c\beta_0x) \end{aligned} \tag{8.6}$$

where

$$g(x) = \begin{cases} \gamma_0, & x \le x_1 \\ < \gamma_0, & x > x_1 \end{cases} \tag{8.7}$$

Here $x^* = m/c\beta_0$ and assume $x^* < x_1$.

Since, in the preceding example, the two predators behave identically, the ratio y_2/y_1 remains constant and the system is as though there were a single predator with population given by $y = y_1 + y_2$, so the system (8.6) is equivalent to

$$x' = xg(x) - \beta_0 xy$$
$$y' = y(-m + c\beta_0 x) \qquad (8.8)$$

For $x \leq x_1$, system (8.8) is a Lotka-Volterra system, and hence there must be a unique periodic orbit Γ_1 tangent to the line $x = x_1$. Inside Γ_1, all solutions are periodic, and, further, all orbits outside of Γ_1 are attracted to Γ_1. Hence there is a disc D_1 with Γ_1 in its interior which is a two-dimensional attractor block. Let $B_1 = \{(x,y_1,y_2) \mid (x,y_1 + y_2) \in D_1\}$. Then B_1 is an attractor block for (8.6). The reason for this is that if (x_0,y_{10},y_{20}) is an initial point so close to B_1 that $\lim_{t\to\infty}[x(t), y_1(t) + y_2(t)] \in D_1$, then by the relation of system (8.8) to (8.6), $\lim_{t\to\infty}[x(t), y_1(t), y_2(t)] \in B_1$.

The second stage is to consider the system

$$x' = xg(x) - y_1 p_1(x) - \beta_0 y_2 x$$
$$y_1' = y_1[-m + cp_1(x)] \qquad (8.9)$$
$$y_2' = y_2(-m + c\beta_0 x)$$

where $g(x)$ is given by (8.7) and $p_1(x)$ satisfies

$$\frac{dp_1(x)}{dx} = \beta_1 < \beta_0 \qquad \text{for all } x \in I_1$$
$$p_1(x^*) = \frac{m}{c} \qquad (8.10)$$

$p_1(x) = \beta_1 x + m/c\ (1 - \beta_1/\beta_0)$ in I_1 and so it is clear that these conditions can be met. Here I_1 is the projection on the x-axis of B_1 (or D_1). Also choose $p_1(x)$ so close to $\beta_0 x$ that B_1 remains an attractor block (small perturbations of systems preserve attractor or repeller blocks).

In the x-y_1 plane, the interior equilibrium is unstable and hence is a repeller. As such there must be a disk D_2 which is a repeller block containing the equilibrium. Let I_2 be the projection of D_2 on the x-axis (see Fig. 8.1a).

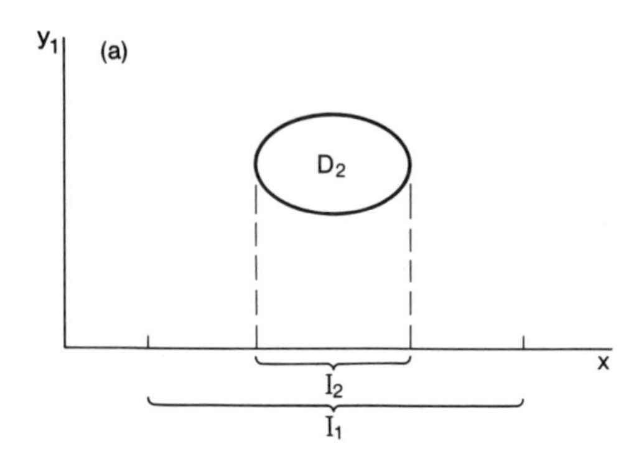

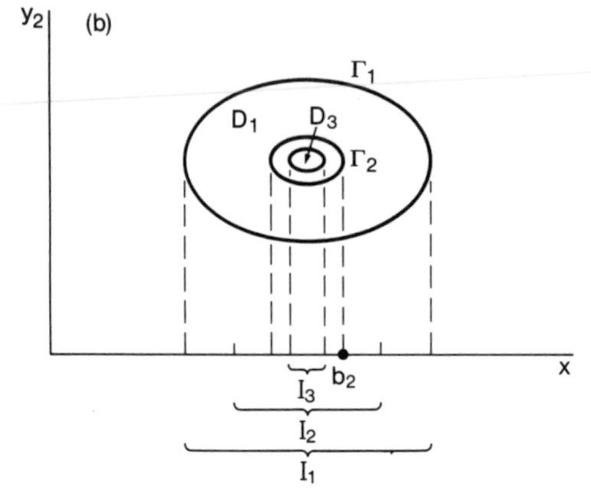

Fig. 8.1 Construction of the intervals (a) I_2, (b) I_3.

Now we come to the third stage. Considering D_1 and Γ_1 as a disk and curve, respectively, in the x-y_2 plane, choose a nontrivial periodic orbit Γ_2 inside Γ_1 and such that the projection of Γ_2 on the x-axis is contained properly within I_2. Let b_2 be the right end point of this projection. Then consider the system

$$\begin{aligned} x' &= xg(x) - y_1p_1(x) - \beta_0 y_2 x \\ y' &= y_1[-m + cp_1(x)] \qquad (8.11) \\ y_2' &= y_2(-m + c\beta_0 x) \end{aligned}$$

where now

$$g(x) = \begin{cases} \gamma_0, & x \leq b_2 \\ < \gamma_0, & x > b_2 \end{cases} \tag{8.12}$$

everything else remaining the same, and where $g(x)$ is so close to the $g(x)$ of (8.7) that B_1 and D_2 continue to remain attractor and repeller blocks, respectively, in their appropriate spaces.

Just as before in the x-y_2 plane we construct an attractor disk D_3 such that its projection on the x-axis I_3 is contained properly within I_2. Now the attractor block B_1 can be shrunk down to an attractor block B_3, such that $B_3 \cap \{x\text{-}y_1 \text{ plane}\} = D_1$, $B_3 \cap \{x\text{-}y_2 \text{ plane}\} = D_3$, and the boundary of B_3 is transverse to the x-y_2 and x-y_1 planes (see Fig. 8.1b) for the construction of I_3).

For the fourth-stage model we use

$$\begin{aligned} x' &= xg(x) - y_1p_1(x) - y_2p_2(x) \\ y_1' &= y_1[-m + cp_1(x)] \\ y_2' &= y_2[-m + cp_2(x)] \end{aligned} \tag{8.13}$$

where $p_2(x)$ satisfies

$$\begin{aligned} \frac{dp_2(x)}{dx} &= \beta_2 < \beta_1, \qquad x \in I_1 \\ p_2(x^*) &= \frac{m}{c} \end{aligned} \tag{8.14}$$

so that for $x \in I_1$, $p_2(x) = \beta_2 x + (m/c)(1 - \beta_2/\beta_0)$, allowing our conditions to be met, and where $p_2(x)$ is chosen so close to $\beta_0 x$ that B_3 and D_2 remain attractor and repeller blocks, respectively, everything else remaining the same.

Since $p_1(x^*) = p_2(x^*)$, there is a line of equilibria given by (8.5). Computing the variational matrix about the points in this line shows that in directions away from the line, it repells. Hence a repeller block C_4 around this line can be constructed. Also, since D_2 is a repeller block in the x-y_1 plane, C_4 can be extended to a repeller block C so that $C \cap \{x\text{-}y \text{ plane}\} = D_2$. Also C can be chosen so its boundary is transverse to both the x-y_1 and x-y_2 planes. Now, letting $B_4 = B_3 \setminus \text{interior}\ (C)$, we find that B_4 is an attractor block. Also let $D_4 = C \cap \{x\text{-}y_2 \text{ plane}\}$.

We have now reached the fifth stage of the example. Here the model is given by

$$\begin{aligned} x' &= xg(x) - y_1 q_1(x) - y_2 p_2(x) \\ y_1' &= y_1[-m + cq_1(x)] \\ y_2' &= y_2[-m + cp_2(x)] \end{aligned} \tag{8.15}$$

where

$$q_1(x) = \begin{cases} p_1(x), & x \in I_3 \\ < p_1(x), & x \in I_1 \setminus \text{interior}\ (I_2) \end{cases} \tag{8.16}$$

and everything else is the same as in (8.13). In addition, $q_1(x)$ is chosen so close to $p_1(x)$ that B_4 remains an attractor block for (8.15).

We now consider the function

$$L_2(x,y_1,y_2) = \frac{y_2^{c\beta_1}}{y_1^{c\beta_2}}$$

Computing dL_2/dt, we get for $x \in I_1$,

$$\frac{dL_2}{dt} = \beta_2 c^2 \frac{y_2^{c\beta_1}}{y_1^{c\beta_2}} \left[\frac{m}{c} + \beta_1(x - x^*) - q_1(x)\right] \tag{8.17}$$

But $q_1(x)$ is chosen so that $m/c + \beta_1(x - x^*) - q_1(x) \geq 0$ for $x \in I_1$ with strict inequality holding for $x \in I_1 \setminus \text{interior}\ (I_2)$. Hence the entire annulus $D_1 \setminus \text{interior}\ (D_2)$ repels in the y_2 direction, which means the attractor block B_4 can be shrunk away from the x-y_1 plane to an attractor block B_5, whose only intersection with the boundaries of the positive octant is in the x-y_2 plane.

Now we come to the sixth and final stage in which the attractor block will be shrunk away from the x-y_2 plane. The final model is

$$\begin{aligned} x' &= xg(x) - y_1q_1(x) - y_2q_2(x) \\ y_1' &= y_1[-m + cq_1(x)] \\ y_2' &= y_2[-m + cq_2(x)] \end{aligned} \tag{8.18}$$

where

$$q_2(x) < p_2(x), \qquad x \in I_3 \setminus \{x^*\} \tag{8.19}$$

everything else being the same as in (8.16) and $q_2(x)$ is chosen so close to $p_2(x)$ that B_5 remains an attractor block for system (8.18).

Now we consider $L_1(x,y_1,y_2) = y_1^{c\beta_2}/y_2^{c\beta_1}$. For $x \in I_3$, computing dL_1/dt, we get

$$\frac{dL_1}{dt} = \beta_1 c^2 \frac{y_1^{\beta c_2}}{y_2^{\beta c_1}} \left[\frac{m}{c} + \beta_2(x - x^*) - q_2(x)\right] \tag{8.20}$$

But $q_2(x)$ is such that (8.20) is positive and hence B_5 can be shrunk away from the x-y_2 plane to give an attractor block B_6 of system (8.18) which lies interior to the positive octant. Hence system (8.18) exhibits persistence. The attractor block B_6 is illustrated in Fig. 8.2.

What has been shown by this example is that it is possible to have a system modeling two distinct predators coexisting on a single prey. This is obtained by virtue of the fact that an attracting block must contain an asymptotically stable orbit. It is not known whether this orbit is a periodic orbit or a nonperiodic recurrent orbit.

As a footnote to this section, it should be mentioned that more recently Zicarelli (1975), utilizing the Hopf bifurcation theorem, has shown how to construct an example of a persistent system simulating n predators competing for one prey.

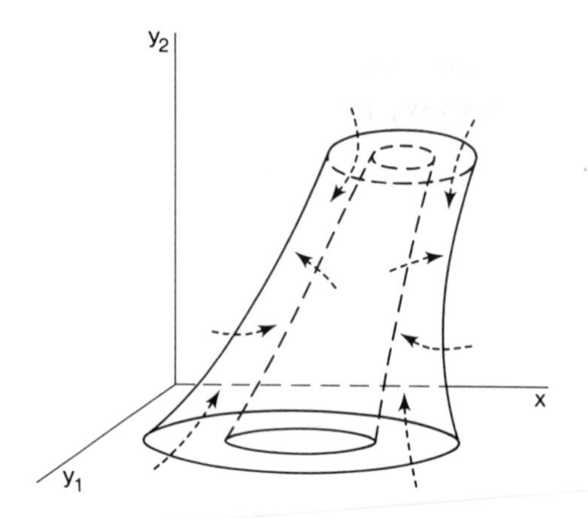

Fig. 8.2 The attractor block B_6.

8.5 KOLMOGOROV-TYPE MODELS: TWO SPECIES

We will consider in this section two species of Kolmogorov-type competition models. For the models we utilize the system

$$\begin{aligned} x_1' &= x_1 f_1(x_1, x_2) \\ x_2' &= x_2 f_2(x_1, x_2) \end{aligned} \tag{8.21}$$

Rescigno and Richardson (1967) have given conditions on system (8.21) so that x_1 and x_2 simulate competitors. In the first instance, increasing the number of either species in f_1 or f_2 should cause a decrease in specific growth rates. This implies that

$$\frac{\partial f_i(x_1, x_2)}{\partial x_j} < 0, \quad i, j = 1, 2 \tag{8.22}$$

Further, in the absence of competitors, both populations should grow if they are small. Hence

$$f_i(0,0) > 0, \quad i = 1, 2 \tag{8.23}$$

One of the effects of competition should be that neither population, even if very small, can grow when in the presence of too large a competing population. In addition, it is supposed that there are carrying capacities for each population. This is tantamount to supposing that there exists A_i, $B_i > 0$, such that

$$f_i(A_i,0) = f_i(0,B_i) = 0, \quad i = 1, 2 \tag{8.24}$$

Consider now the x_1-x_2 plane. In Fig. 8.3(a) and (b) we have drawn the isoclines $f_i(x_1,x_2) = 0$, $i = 1, 2$. Figure 8.3(a) illustrates the case where the isoclines do not intersect, and Fig. 8.3(b) illustrates the case where they do. In each case the quadrant is divided into three zones, I: the area "below" the isoclines, II: the area "between" the isoclines (and including the isoclines), III: the area "above" the isoclines.

Since in zone I, $f_i > 0$, $i = 1, 2$, solutions initiating in zone I increase and approach zone II. Also, since in zone III, $f_i < 0$, $i = 1, 2$, solutions initiating in zone III decrease and approach zone II. Hence either all solutions enter zone II at a finite time or the ω-limit set of the solution is at an equilibrium (at an intersection of the isoclines or on one of the axes).

In the case where the isoclines do not intersect, the only equilibria are on the axes, and since solutions in zone II are monotonic, the ω-limit set must be an equilibrium and hence one species goes extinct. In the case where the isoclines do intersect, coexistence could occur. Solutions could approach any of the interior equilibria, or extinction could occur. The stability of each equilibrium would have to be determined to analyze which one of these happens.

The foregoing has been properly formulated and rigorously proved by Albrecht et al. (1974).

The book by Hirsch and Smale (1974) has a nice geometrical analysis in Chap. 12 of a Kolmogorov-type system, also adapted from Rescigno and Richardson (1967).

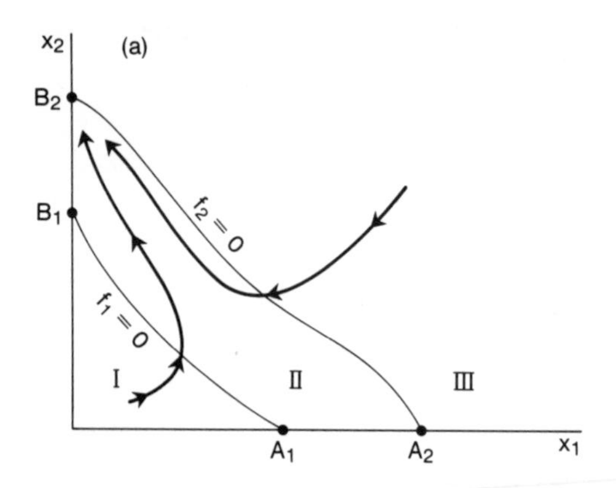

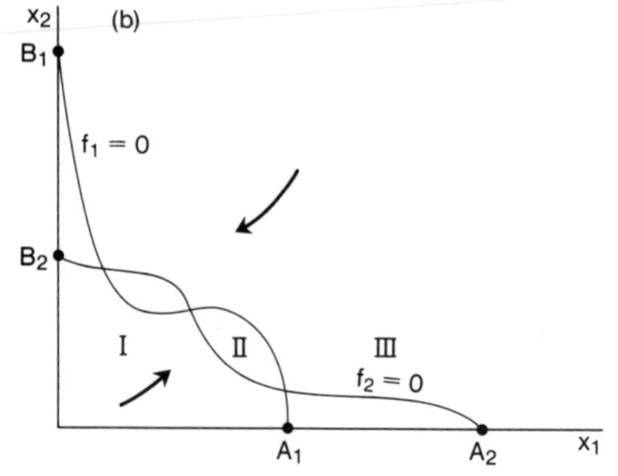

Fig. 8.3 Phase diagram for two-dimensional Kolmogorov competition model. (a) Isoclines not intersecting. (b) Isoclines intersecting.

8.6 KOLMOGOROV-TYPE MODELS: THREE SPECIES

The following model is due to Rescigno (1968).

$$\begin{aligned} x_1' &= x_1 f_1(x_1, x_2, x_3) \\ x_2' &= x_2 f_2(x_1, x_2, x_3) \\ x_3' &= x_3 f_3(x_1, x_2, x_3) \end{aligned} \tag{8.25}$$

For reasons similar to the two-dimensional case, the following assumptions are made:

$$\frac{\partial f_i}{\partial x_j}(x_1,x_2,x_3) < 0, \quad f_i(0,0,0) > 0, \quad i, j = 1, 2, 3 \tag{8.26}$$

$$\begin{aligned} &\exists u_1(x_2,x_3),\ u_2(x_1,x_3),\ u_3(x_1,x_2) \ni \\ &\quad u_1(x_{20},x_{30}) = 0 \Rightarrow f_1(0,x_{20},x_{30}) = 0 \\ &\quad u_2(x_{10},x_{30}) = 0 \Rightarrow f_2(x_{10},0,x_{30}) = 0 \\ &\quad u_3(x_{10},x_{20}) = 0 \Rightarrow f_3(x_{10},x_{20},0) = 0 \end{aligned} \tag{8.27}$$

$$\exists A,B,C > 0 \ni f_1(A,0,0) = f_2(0,B,0) = f_3(0,0,C) = 0 \tag{8.28}$$

Again the phase space can be naturally divided into three zones, separated by the isocline surfaces, $f_i = 0$, $i = 1, 2, 3$. Zone I in the positive octant will consist of those points at which $f_i > 0$, $i = 1, 2, 3$; zone III will be where $f_i < 0$, $i = 1, 2, 3$; and zone II will be all other points of the octant (i.e., the points between and on the isoclines). As before, it is easily seen that solutions initiating in zones I and III either enter zone II in finite time or approach one of the coordinate planes.

Solutions in zone II can have various behaviors depending on how the isoclines intersect. Rescigno (1968) has described the various possibilities. There is as yet no rigorous formulation and mathematical analysis of theorems involving system (8.25)

NOTES ON THE LITERATURE

There have been several papers not previously mentioned which also deal with models having nonlinear specific growth rates and the question of persistence and extinction. These include Armstrong and McGehee (1976a,b), Koch (1974b) and Smith et al. (1975).

There have been discrete competition models as well. See, for example, Hassell and Comins (1976), Haussmann (1971), and Shapiro. Leslie and Gower (1958) and Ludwig (1975) have looked at stochastic models, Levin (1974) has considered a competition model with dispersion, and Albrecht et al. (1976) have discussed models with controls.

The following papers contain results on competition models in spatially heterogeneous environments: Culver (1973) and Lawlor and Maynard Smith (1976). Some papers involving competition models for species in a chemostat are S. B. Hsu (1976) and Hsu et al. (1977, (1978).

Some further papers written on exclusion and coexistence are Haigh and Maynard Smith (1972), Haussmann (1973), Phillips (1973), Rescigno and Richardson (1965), and Smale (1976).

Other papers dealing with various aspects of competition theory are Cunningham (1955), Gilpin (1974a), Gilpin and Justice (1973), Leon and Tumpson (1975), Miller (1969), Neill (1974), Schoener (1974a), Stewart and Levin (1973), Utz (1961), Wangersky and Cunningham (1957), and Wiegert (1974).

EXERCISES

8.1 (S. B. Hsu, personal communication) Consider the system

$$x'(t) = [x^{(0)} - x(t)]D - \sum_{i=1}^{n} \frac{m_i}{c_i} \frac{y_i(t)x(t)}{a_i + x(t)}$$

$$y'(t) = \frac{m_i y_i(t)x(t)}{a_i + x(t)} - Dy_i(t), \qquad i = 1, \ldots, n$$

$$x(0) = x_0 > 0, \qquad y_i(0) = y_{i0} > 0$$

(a) Show that if $m_i/D \leq 1$, $i = 1, \ldots, n$ or $a_iD/(m_i - D) > x^{(0)}$ if $m_{i/D} > 1$, then $\lim_{t\to\infty} y_i(t) = 0$.

(b) Show that if

$$0 < \frac{a_1D}{m_1 - D} < \frac{a_2D}{m_2 - D} \leq \cdots \leq \frac{a_nD}{m_n - D} \qquad \text{and}$$

$$\frac{a_1D}{m_1 - D} < x^{(0)} \text{ then } \lim_{t\to\infty} x(t) = \frac{a_1D}{m_1 - D},$$

$$\lim_{t\to\infty} y_1(t) = c_1\left(x_0 - \frac{a_1D}{m_1 - D}\right), \qquad \lim_{t\to\infty} y_i(t) = 0,$$

$2 \leq i \leq n$

8.2 (Open) For system (8.3) derive general criteria for exclusion to hold and general cirteria which guarantee persistence.

8.3 (Open) Analyze the n-dimensional Kolmogorov model

$$x_i' = x_i f_i(x_1, \ldots, x_n).$$

Derive conditions for this model to simulate competition. Derive conditions for equilibria and determine their stability. When do periodic solutions exist?

8.4 (Open problem suggested by S. B. Hsu) Analyze the two-resources, two-predator system

$$S' = (S^{(0)} - S)D_S - \frac{k_1 x_1 S}{a_1 + \omega_1 S + \omega_2 R} - \frac{k_2 x_2 S}{a_2 + \omega_3 S + \omega_4 R}$$

$$R' = (R^{(0)} - R)D_R - \frac{k_3 x_1 R}{a_1 + \omega_1 S + \omega_2 R} - \frac{k_4 x_2 R}{a_2 + \omega_3 S + \omega_4 R}$$

$$x_1' = \left(\frac{b_1 S + b_2 R}{a_1 + \omega_1 S + \omega_2 R} - D_1\right)x_1$$

$$x_2' = \left(\frac{b_3 S + b_4 R}{a_2 + \omega_3 S + \omega_4 R} - D_2\right)x_2$$

Chapter 9
COOPERATION (SYMBIOSIS)

9.1 INTRODUCTION

Very little work has been done mathematically on models of cooperation. In this chapter we present known results for Lotka-Volterra-type and Kolmogorov-type models. We are not aware of any intermediate-type symbiotic models to date.

9.2 LOTKA-VOLTERRA MODELS

Here we consider two Lotka-Volterra models of cooperation, the first without carrying capacities and the second with carrying capacities.

We first consider the model

$$\begin{aligned} x_1' &= x_1(\alpha_1 + \beta_1 x_2) \\ x_2' &= x_2(\alpha_2 + \beta_2 x_1) \end{aligned} \tag{9.1}$$

Since we clearly have $x_i' > \alpha_i x_i$, then $x_i > x_{i0} e^{\alpha_i t}$ and hence $\lim_{t\to\infty} x_i(t) = +\infty$, $i = 1, 2$. By separation of variables system (9.1) can be solved in the phase plane, the solution of which is

$$\alpha_1 \ln\left(\frac{x_2}{x_{20}}\right) + \beta_1(x_2 - x_{20}) = \alpha_2 \ln\left(\frac{x_1}{x_{10}}\right) + \beta_2(x_1 - x_{10}) \tag{9.2}$$

where (x_{10}, x_{20}) are initial populations. There is, of course, only one equilibrium in the first quadrant, located at the origin, and it is unstable.

A more interesting model is given by the system

$$x_1' = x_1\left(\alpha_1 - \frac{\alpha_1 x_1}{K_1} + \beta_1 x_2\right)$$
$$x_2' = x_2\left(\alpha_2 + \beta_2 x_1 - \frac{\alpha_2 x_2}{K_2}\right) \tag{9.3}$$

Let Δ be given by

$$\Delta = \alpha_1\alpha_2 - \beta_1\beta_2 K_1 K_2 \tag{9.4}$$

There are at least three equilibria in the first quadrant, (E1): (0,0), (E2): $(K_1,0)$, and (E3): $(0,K_2)$. Further, if $\Delta > 0$, there is a fourth such equilibrium (E4): (x_1^*,x_2^*), where

$$x_1^* = \alpha_2 K_1(\alpha_1 + \beta_1 K_2)\Delta^{-1}, \qquad x_2^* = \alpha_1 K_2(\alpha_2 + \beta_2 K_1)\Delta^{-1} \tag{9.5}$$

There are in either case two isoclines, the one given by the equation $\alpha_1 x_1 - \beta_1 K_1 x_2 = \alpha_1 K_1$ and passing through (E2), the other given by $K_2\beta_2 x_1 - \alpha_2 x_2 = -\alpha_2 K_2$ and passing through (E3). Further, (E1) is unstable, (E2) and (E3) are hyperbolic, attracting on the axes and repelling away from the axes. We now consider two cases.

CASE A. $\Delta \le 0$. In this case (E4) does not exist in the positive quadrant and the isoclines do not intersect. The phase plane may be divided into three zones as shown in Fig. 9.1a. All solutions initiating in zones I and III must cross into zone II in finite time. All solutions in zone II remain there and grow unboundedly (see Fig. 9.1a).

CASE B. $\Delta > 0$. In this case (E4) exists and is given by (9.5). If we compute the variational matrix V at (E4), we get

$$V = \begin{bmatrix} \dfrac{-\alpha_1 x_1^*}{K_1} & \beta_1 x_1^* \\ \\ \beta_2 x_2^* & \dfrac{-\alpha_2 x_2^*}{K_2} \end{bmatrix} \tag{9.6}$$

which has the characteristic polynomial

$$\lambda^2 + \left(\frac{\alpha_1 x_1^*}{K_1} + \frac{\alpha_2 x_2^*}{K_2}\right)\lambda + \frac{x_1^* x_2^*}{K_1 K_2}\Delta \tag{9.7}$$

From (9.7), the eigenvalues are given by

$$\lambda = -\frac{1}{2}\left(\frac{\alpha_1 x_1^*}{K_1} + \frac{\alpha_2 x_2^*}{K_2}\right) \pm \frac{1}{2}\left[\left(\frac{\alpha_1 x_1^*}{K_1} + \frac{\alpha_2 x_2^*}{K_2}\right)^2 - 4\,\frac{x_1^* x_2^*}{K_1 K_2}\Delta\right]^{\frac{1}{2}} \tag{9.8}$$

and since $\Delta > 0$, Re $\lambda < 0$, which gives that (E4) is asymptotically stable. See Fig. 9.1b for the dynamics of this case.

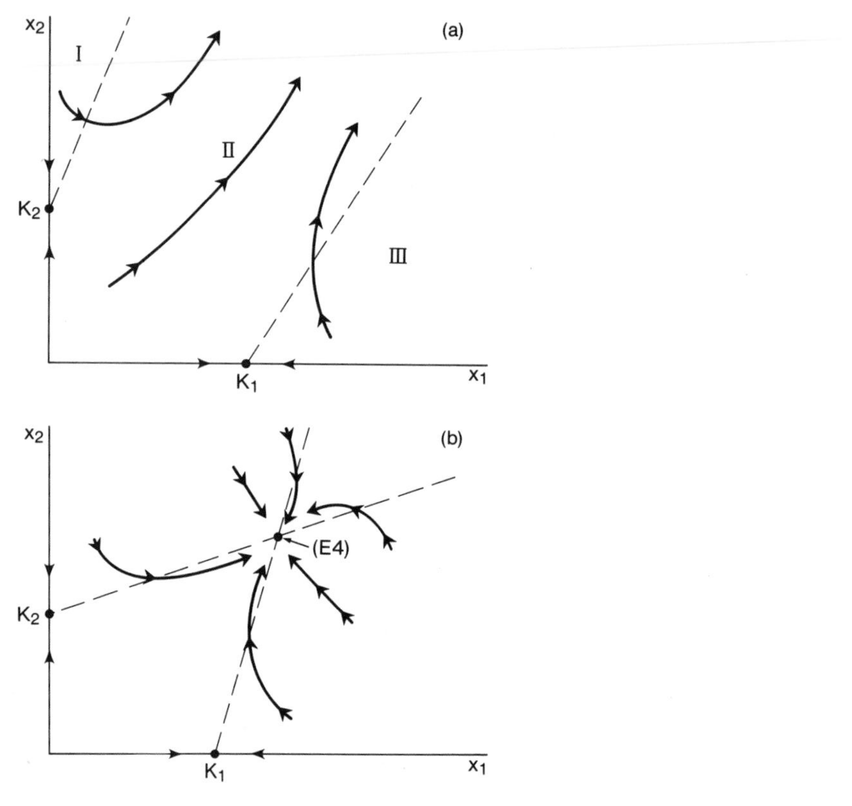

Fig. 9.1 The phase plane dynamics of system (9.3). (a) $\Delta \leq 0$, (b) $\Delta > 0$.

9.3 KOLMOGOROV-TYPE MODELS

In this section we once more consider a Kolmogorov-type growth model

$$\begin{aligned} x_1' &= x_1 f_1(x_1, x_2) \\ x_2' &= x_2 f_2(x_1, x_2) \end{aligned} \tag{9.9}$$

where the $f_i(x_1,x_2)$ have continuous first derivatives throughout the first quadrant (including axes). The assumptions simulating cooperation, given by Rescigno and Richardson (1967), are as follows:

ASSUMPTION A. The growth of each species is enhanced by the presence of the other. Hence

$$\frac{\partial f_1}{\partial x_2} > 0, \qquad \frac{\partial f_2}{\partial x_1} > 0, \qquad x_i \geq 0, \quad i = 1, 2 \tag{9.10}$$

ASSUMPTION B. For a fixed ratio of x_1/x_2, the growth rate is hindered by an increase in the value of both populations, and hence

$$x_1 \frac{\partial f_i}{\partial x_1} + x_2 \frac{\partial f_i}{\partial x_2} \leq -\alpha < 0, \qquad x_i \geq 0, \quad i = 1, 2 \tag{9.11}$$

where α is some positive constant.

ASSUMPTION C. If both populations are very small, then they grow, or

$$f_i(0,0) > 0, \qquad i = 1, 2 \tag{9.12}$$

ASSUMPTION D. Finally, there is a carrying capacity for each population, i.e., there are positive constants K_1, K_2 such that

$$f_1(K_1,0) = f_2(0,K_2) = 0 \tag{9.13}$$

Under the preceding assumptions one can show the existence of an interior equilibrium (E). Let the coordinates of (E) be (x_1^*, x_2^*). Then the variational matrix of system (9.9) about (E), V, is

$$V = \begin{bmatrix} x_1^* \frac{\partial f_1}{\partial x_1}(x_1^*,x_2^*) & x_1^* \frac{\partial f_1}{\partial x_2}(x_1^*,x_2^*) \\ x_2^* \frac{\partial f_2}{\partial x_1}(x_1^*,x_2^*) & x_2^* \frac{\partial f_2}{\partial x_2}(x_1^*,x_2^*) \end{bmatrix} \tag{9.14}$$

Clearly the trace of V is negative by (9.10) and (9.11). Also from (9.11),

$$\frac{\partial f_1}{\partial x_1} \leq \frac{-\alpha}{x_1} - \frac{x_2}{x_1}\frac{\partial f_1}{\partial x_2} \qquad \text{and} \qquad \frac{\partial f_2}{\partial x_2} \leq \frac{-\alpha}{x_2} - \frac{x_1}{x_2}\frac{\partial f_2}{\partial x_1}$$

whence

$$\frac{\partial f_1}{\partial x_1}\frac{\partial f_2}{\partial x_2} - \frac{\partial f_1}{\partial x_2}\frac{\partial f_2}{\partial x_1} \geq \frac{\alpha^2}{x_1x_2} + \alpha\left(\frac{1}{x_1}\frac{\partial f_2}{\partial x_1} + \frac{1}{x_2}\frac{\partial f_1}{\partial x_2}\right) > 0$$

for x_1, $x_2 > 0$

giving that the determinant

$$|V| = x_1^*x_2^*\left[\frac{\partial f_1}{\partial x_1}\frac{\partial f_2}{\partial x_2} - \frac{\partial f_1}{\partial x_2}\frac{\partial f_2}{\partial x_1}\right](x_1^*,x_2^*) \tag{9.15}$$

must be positive. Hence the characteristic polynomial

$$\lambda^2 - (\text{tr } V)\lambda + |V| \tag{9.16}$$

has roots with negative real parts and is asymptotically stable.

It has been shown rigourously by Albrecht et al. (1974) that (E4) is globally asymptotically stable in the first quadrant. The phase portrait of system (9.9) is similar to that in Fig. 9.1b (the isoclines are in general here not straight lines).

NOTES ON THE LITERATURE

In addition to the two previously mentioned papers we also mention the work in Albrecht et al. (1976), which deals with symbiotic systems with controls, and the work in Roughgarden (1975d), which deals with the question of whether or not a guest derives cost benefits for searching out certain species of hosts.

EXERCISES

9.1 (a) Analyze an n-dimensional Lotka-Volterra system for cooperation.

(b) For an n-dimensional Kolmogorov model, set down appropriate conditions simulating symbiosis and analyze the model.

APPENDIX

In this appendix various aspects of analysis and differential equations are given, which are crucial to the understanding of the material in this text. Techniques are shown when necessary, but proofs of theorems are not given, although references to where the proofs may be found are included.

A: PERTURBATION THEORY

Perturbation theory is used for equations or systems of equations which are "close" in some sense to other equations or systems that can be analyzed. For example, if u is a function in some functional space and L is a linear operator on that space such that the equation

$$Lu = 0 \tag{A.1}$$

can be solved, then perturbational techniques might yield information about solutions of the equation

$$(L + M)u = 0 \tag{A.2}$$

provided that the operator L + M is in some sense sufficiently close to the operator L.

As it is utilized in this book perturbation theory is applied to systems of differential equations. The perturbations are described by means of a parameter (usually ε) and closeness of two systems corresponds to ε being sufficiently small.

The underlying assumption then becomes that if ε is sufficiently small, solutions of the perturbed system behave in a manner close to the behavior of the unperturbed system. One of the most common techniques in analyzing this behavior is to assume the solutions may be written in a power series expansion in ε (sometimes in fractional powers of ε) and to separate out the coefficients of the powers of ε, solving each equation successively.

For example, suppose that Eq. (A.2) is written

$$(L + \varepsilon L_1)u = 0 \tag{A.3}$$

Then assuming $u = u_0 + \varepsilon u_1 + \varepsilon^2 u_2 + \ldots$, substitution into (A.3) and separation of powers of ε yields the system of equations

$$\begin{aligned} Lu_0 &= 0 \\ Lu_1 + L_1 u_0 &= 0 \\ Lu_2 + L_1 u_1 &= 0 \\ &\vdots \end{aligned} \tag{A.4}$$

The first equation is just (A.1), which is presumably solved. Then if a technique for solving the second equation of (A.4) can be found, u_1 is known. The same technique probably solves the remaining equations in (A.4) successively so as to give the solution to any desired accuracy.

For a nice treatise on perturbation theory, especially for ordinary differential equations, see Bellman (1953).

B: THE IMPLICIT FUNCTION THEOREM

The implicit function theorem has to do with the question of when in a given relation between variables some of the variables can be solved in terms of the others. In this text, as the theorem is used, all variables are real, and the relation is usually in terms of a system of equations.

B.1 THE SCALAR, NONCRITICAL CASE

Let $f(x,y)$ be a real function of the real variables x and y, and suppose a and b are such that

$$f(a,b) = 0 \tag{A.5}$$

The question of interest is when the equation

$$f(x,y) = 0 \tag{A.6}$$

can be solved for y in terms of x in a neighborhood of (a,b). The first criterion does not presume any differentiability conditions, but it does assume monotonicity conditions.

THEOREM A.1 *Let* $f(x,y)$ *be continuous in* x *and* y *and monotonic in the* y *variable in a neighborhood of* (a,b). *Then* (A.6) *can be solved for* $y = \phi(x)$, $\phi(a) = b$, *such that* $f(x,\phi(x)) = 0$ *in a neighborhood of* (a,b).

The most common form of the implicit function theorem assumes some differentiability.

THEOREM A.2 *Let* $f(x,y)$ *be continous in a neighborhood of* (a,b).

(i) *If* $\partial f(a,b)/\partial y \neq 0$, *then* (A.6) *can be solved for* y *as a function of* x, $y = \phi(x)$, *such that* $\phi(a) = b$, *and such that* $f(x,\phi(x)) = 0$ *in a neighborhood of* $x = a$.

(ii) *If, in addition,* $\partial f/\partial y$ *is continuous in a neighborhood of* (a,b), *then* $\phi(x)$ *is continuous in a neighborhood of* $x = a$.

(iii) *If also* $(\partial f/\partial x)(a,b)$ *exists, then* $\phi'(a)$ *exists and* $\phi'(a) = -(\partial f/\partial x)(a,b)/(\partial f/\partial y)(a,b)$. *In addition, if* $\partial f/\partial x$ *is continuous, so is* $\phi'(x)$. *In this case* $\phi(x) = b + \phi'(a)(x - a) + o(x - a)$.

The reader is referred to Rudin (1953) for a proof the theorem in the noncritical case.

B.2 SCALAR CRITICAL CASES

At this time the critical case where $(\partial f/\partial y)(a,b) = 0$ is considered. There are two critical cases of importance in this book.

In the first instance suppose that as well

$$\frac{\partial f}{\partial x}(a,b) = 0 \tag{A.7}$$

Then letting $y = b + \alpha(x - a)$, $g(x,\alpha) = (x - a)^{-2} f[x, b + \alpha(x - a)]$, and applying L'Hospital's rule, it is clear that

$$g(a,\alpha) = \frac{1}{2} f_{yy}(a,b)\alpha^2 + f_{xy}(a,b)\alpha + \frac{1}{2} f_{xx}(a,b) \tag{A.8}$$

and

$$g_\alpha(a,\alpha) = f_{yy}(a,b)\alpha + f_{xy}(a,b) \tag{A.9}$$

The condition that we can solve $g(x,\alpha) = 0$ for α as a function of x such that $g(a,0) = 0$ is that $g(a,\alpha)$ as given by (A.8) has a real root α_0 such that $g_\alpha(a,\alpha_0) \neq 0$. This is clearly guaranteed if the discriminant

$$f_{xy}(a,b)^2 - f_{yy}(a,b) f_{xx}(a,b) > 0 \tag{A.10}$$

Suppose now (A.10) holds. Then $f(x,y) = 0$ can be solved for y as a function of x, which is of the form

$$y = b + \alpha_0(x - a) - \frac{g_x(a,\alpha_0)}{g\ (a,\alpha_0)}(x - a)^2 + o(x - a)^2$$

Now we consider the case where (A.7) does not hold. We suppose then for the sake of argument that

$$f_x(a,b) f_{yy}(a,b) < 0 \tag{A.11}$$

and we let $x - a = t^2$. We define

$$E(t,y) = f(a + t^2, y) \tag{A.12}$$

It is easily seen from (A.12), since $E_t(t,y) = 2f_x(a + t^2, y)t$, that

$$E(0,b) = E_t(0,b) = E_y(0,b) = 0 \tag{A.13}$$

Hence, from the previous case, if we let $y = b + \theta t$ and define $D(t,\theta) = t^{-2}E(t,b + \theta t)$, then the condition for $D(t,\theta) = 0$ to be solved for θ as a function of t near (0,0), i.e., $D_{t\theta}^2(0,0) - D_{tt}(0,0)D_{\theta\theta}(0,0) > 0$, is automatically fulfilled, since $D_{\theta\theta}(0,0) = f_{yy}(a,b)$, $D_{t\theta}(0,0) = 0$, and $D_{tt}(0,0) = 2f_x(a,b)$ and (A.11) holds.

In this case the solution is of the form $\theta = \theta_0 - D_t(0,\theta_0)D_\theta(0,\theta_0)^{-1}t + o(t)$, where θ_0 is either of

$$\theta_0 = \pm\left(-\frac{2f_x(a,b)}{f_{yy}(a,b)}\right)^{\frac{1}{2}} \tag{A.14}$$

which gives

$$y = \phi(x) = b + \theta_0(x - a)^{\frac{1}{2}} - D_t(0,\theta_0)D\ (0,\theta_0)^{-1}(x - a) + o(x - a) \tag{A.15}$$

as the solution of $f(x,y) = 0$ near (a,b) valid for positive x - a.

If the opposite inequality to (A.11) holds, letting $x = a - t^2$ will give similar results with the minus sign inside the bracket in (A.14) removed and valid for negative x - a.

For other critical cases of the scalar implicit function theorem see Freedman (1969) and Butler and Freedman (1972).

B.3 THE VECTOR CASE

The case when x and y are m- and n-dimensional vectors, respectively, corresponding to the noncritical case may be stated as follows:

Suppose $f(x,y)$ *is an* n *dimensional vector,* $f(a,b) = 0$, $\det \mid f_y(a,b) \mid \neq 0$, *where* $f_y(a,b)$ *is that* $n \times n$ *matrix whose* ij-*th component is* $\partial f_i/\partial y_i$. *Then* $f(x,y) = 0$ *may be solved for* y *as a function of* x, $y = \phi(x)$, $\phi(a) = b$, *such that* $f(x,\phi(x)) = 0$ *in a neighborhood of* (a,b). *If, further,* $f_x(a,b)$ *exists and* f_y *and* f_x *are continuous near* (a,b), *then*

$$\phi(x) = b + f_y(a,b)^{-1}f_x(a,b)(x - a) + o\|x - a\|$$

Note that f_x is an $n \times n$ matrix. For a proof of this theorem see Rudin (1953). Critical cases are similar to the scalar situation. However, there is an intermediate step in order to sort out the critical rows and columns from the noncritical in the matrix $f_y(a,b)$. See Coddington and Levinson (1955), Chap. 14, for details.

C: EXISTENCE AND UNIQUENESS OF ORDINARY DIFFERENTIAL EQUATIONS

Throughout this text it is always assumed that solutions to initial value problems always exist and are uniqe. In this appendix we present sufficiency theorems guaranteeing these assumptions. See Coddington and Levinson (1973) or Hartman (1976a) for proofs and further details.

C.1 EXISTENCE

The question of existence relevant to this text has to do with the so-called initial value problems of ordinary differential equations. Suppose we consider the system

$$x' = f(t,x) \tag{A.16}$$

where x is an n-dimensional vector, t a scalar, and f an n-dimensional vector each component of which is a function of n + 1 variables (t and the n components of x). We wish to consider solutions of (A.16) which satisfy the initial condition

$$x(t_0) = x_0 \tag{A.17}$$

where x_0 is a specified n-dimensional vector, and the question of whether or not solutions of (A.16) subject to (A.17) exist is the question of interest.

The following known as the Cauchy-Peano existence theorem, is the most relevant existence theorem for our purposes.

Let R *be the rectangle,* R: $|t - t_0| \le a$, $\|x - x_0\| \le b$ $(a,b > 0)$. *Let* f *be continuous over* R *and let* $M = \max \|f(t,x)\|$, $[(t,x) \in R]$ *and let* $\alpha = \min(a, b/M)$. *Then there exists a differentiable function*

$\phi(t)$ *which is a solution of* (A.16) *on the inverval* $|t - t_0| \leq \alpha$ *and which satisfies* (A.17).

See Coddington and Levinson (1955) for a proof of this theorem.

In connection with the question of existence there occurs the question of extendibility (i.e., the question of when a larger value of α can be used). The reader is referred to Hartman (1973) for extendibility theorems and for more general existence theorems.

C.2 UNIQUENESS

For purposes of modeling in mathematical ecology, we want our models to have unique solutions. For this purpose we define what it means for a function $f(x)$ to possess the Lipschitz property.

Let $f(x)$ be a scalar function defined in some interval d. Then $f \in \text{Lip}\ (d)$ if there exists $k > 0$ such that $|f(x_2) - f(x_1)| \leq k|x_2 - x_1|$ for all $x_1, x_2 \in d$.

Note that a differentiable function on d is automatically Lipschitzian and a Lipschitz function is automatically continuous. The following is the most important uniqueness theorem for our purposes (here f and x are again vectors).

Let R, M, *and* α *be as in Sec.* C.1. *Let for each fixed* x, $f(t,x)$ *be continuous in* t *and for each fixed* t, $\|f(t,x)\|$ *be Lipschitzian in* $\|x\|$, $(t,x) \in R$. *Then there is a unique, continuously differentiable solution of* (A.16) *on the interval* $|t - t_0| < \alpha$ *and which satisfies* (A.17).

Again for a proof see Coddington and Levinson (1955). Also, the property of $\|f(t,x)\|$ being Lipschitzian in $\|x\|$ will be automatically fulfilled if each component of $f(t,x)$ possesses bounded partial derivatives with respect to the components of x.

D: STABILITY AND PERIODICITY

A great deal of time in this text is spent showing the existence of equilibria and periodic solutions and in the analysis of their stability. In this section of the appendix we state criteria for the stability of equilibria of first linear and then nonlinear systems. We also look at the stability of periodic solutions. The topic of Liapunov functions is also briefly discussed for the purpose of determining both stability and periodicity criteria for autonomous systems. All material, except the material on Liapunov functions, together with proofs may be found in Coddington and Levinson (1955). The material on Liapunov functions may be found in La Salle and Lefschetz (1961).

D.1 LINEAR EQUATIONS: CONSTANT COEFFICIENTS

The general linear system with constant coefficients may be written in the form

$$x' = Ax \qquad (' = \frac{d}{dt}) \tag{A.18}$$

where x is the n-dimensional vector whose i-th component is x_i and A is the $n \times n$ matrix whose ij-th component is a_{ij}.

The solution of (A.18) subject to the initial condition (A.17) is

$$x = e^{A(t - t_0)} x_0 \tag{A.19}$$

where by e^M, M being an $n \times n$ matrix, we mean the convergent series

$$e^M = \sum_{k=0}^{\infty} \frac{M^k}{k!} \tag{A.20}$$

At this time we will be interested in the stability of the zero solution ($x_0 = 0$). $x = 0$ is said to be *stable* if given any $\varepsilon > 0$, there is a $\delta > 0$ such that all solutions $x(t)$ for which $\|x(t_0)\| < \delta$ for some t_0, then $\|x(t)\| < \varepsilon$ for $t \geq t_0$. In addition, if $\lim_t x(t) = 0$, then $x = 0$ is said to be *asymptotically stable*. If $x = 0$ is not stable, it is said to be *unstable*.

The following criteria tell the complete story for system (A.18).

If all eigenvalues of A *have negative real parts,* $x = 0$ *is asymptotically stable. If one eigenvalue of* A *has a positive real part,* $x = 0$ *is unstable. If no eigenvalues have positive real parts, but one or more have zero real parts, then if the eigenvalues with zero real parts have simple elementary divisors (i.e., there are as many independent eigenvectors belonging to each of those eigenvalues as their respective multiplicities),* $x = 0$ *is stable, but if any of the eigenvalues with zero real part do not have simple elementary divisors,* $x = 0$ *is unstable.*

D.2 LINEAR EQUATIONS: PERIODIC COEFFICIENTS

Now we are interested in the stability of the zero solution of

$$x' = A(t)x \tag{A.21}$$

where $A(t + \omega) = A(t)$ for all t.

Let $\Phi(t)$ be an $n \times n$ matrix, each column of which is an independent solution of (A.21). $\Phi(t)$ is called a fundamental matrix of (A.21).

In the case where A(t) is periodic, Floquet's theorem states that there is a periodic matrix $P(t) = P(t + \omega)$ and a constant matrix R such that

$$\Phi(t) = P(t)e^{Rt}$$

If R is known, then so is the stability of $x = 0$, for the stability of $x = 0$ is determined by the eigenvalues of R in a completely analogous manner to the way in which the eigenvalues of A determine the stability in the constant coefficient case. Hence if the eigenvalues of R have negative real parts, $x = 0$ is asymptotically stable, whereas if one eigenvalue has a positive real part, $x = 0$ is unstable. Again if all real parts are zero or negative, $x = 0$ could be stable or unstable.

Note that the eigenvalues of e^{Rt} are less than 1, equal to 1, or greater than 1 in absolute value corresponding to the eigenvalues of R having negative real part, zero real part, or positive real part. The eigenvalues of e^{Rt} are called the characteristic multipliers of system (A.21).

D.3 THE VARIATIONAL EQUATION

Consider now an autonomous system of differential equations

$$x' = f(x) \tag{A.22}$$

and let $\phi(t)$ be a solution of this system, i.e., $\phi'(t) = f(\phi(t))$. The variational equation of system (A.22) with respect to $\phi(t)$ is the linear part of the expansion of system (A.22) about $\phi(t)$. It is formally given by the linear system

$$y' = f_x(\phi(t))y \tag{A.23}$$

where the variational matrix $f_x(\phi(t))$ is that matrix whose ij-th component is $(\partial f_i/\partial x_j)(\phi(t))$.

Under certain circumstances the stability of $y = 0$ of (A.23) determines the stability of $x = \phi(t)$ of (A.22). Some of these circumstances are discussed in the next two sections.

D.4 STABILITY OF EQUILIBRIA

In the special case that $\phi(t) \equiv \phi_0$, a constant, then if we set $A = f_x(\phi_0)$, system (A.23) reduces to system (A.18) with x replaced by y. Since system (A.23) can be regarded as a small perturbation of system (A.22), it can be shown that if the eigenvalues of A all have negative real parts, then both $y = 0$ and $x = \phi_0$ are asymptotically stable solutions of (A.23) and (A.22), respectively. If one of the eigenvalues of A has a positive real part, then both of the preceding solutions are unstable.

In the case that one or more of the eigenvalues of A has zero real parts (and the rest have negative real parts), the analysis is considerably more complicated. The reader is referred to Coddington and Levinson (1955), Bellman (1953), La Salle and Lefschetz (1961), and Nemytskii and Stepanov (1960) for discussions on this case.

D.5 STABILITY OF PERIODIC SOLUTIONS

Here $\phi(t)$ is periodic, i.e., $\phi(t + \omega) = \phi(t)$. If we let $A(t) = f_x(\phi(t))$, system (A.23) is the same as system (A.21) with x replaced by y. In this case one of the characteristic multipliers is always equal to 1. However, it can be shown that if the other characteristic multipliers are less than 1 in absolute value, $x = \phi(t)$ is asymptotically stable. If one of the other characteristic multipliers is greater than 1 in absolute value, $x = \phi(t)$ is unstable. Again the critical case (other characteristic multipliers equal to 1 in absolute value)needs further analysis.

D.6 LIAPUNOV FUNCTIONS

Let $V(x)$ be a scalar function of the n components of x. By $\dot{V}(x)$, the derivative of V along solutions of system (A.22), we mean

$$\dot{V}(x) = \sum_{i=1}^{n} \frac{\partial V(x)}{\partial x_i} f_i(x) \tag{A.24}$$

$\dot{V}(x)$ represents how V changes (increases or decreases) along solutions.

$V(x)$ is said to be a positive definite function if $V(0) = 0$, and outside the origin, but inside some domain D containing the origin, $V(x) > 0$.

Suppose now that system (A.22) has an equilibrium at $x = 0$, i.e., $f(0) = 0$ [if $f(x_0) = 0$, $x_0 \neq 0$, the change of variables $y = x - x_0$ gives a new system with an equilibrium at the origin]. The following results, which are presented and proved in La Salle and Lefschetz (1961), give stability criteria for $x = 0$ of system (A.22).

Let $V(x)$ be a positive definite function in D. Let $\Omega \subset D$ be a subregion of D containing the origin in its interior.

1. If $\dot{V}(x) \leq 0$, $x \in \Omega$, then $x = 0$ is stable.
2. If $\dot{V}(x) < 0$, $x \in \Omega$, then $x = 0$ is asymptotically stable and the region of attraction contains Ω.
3. If $\dot{V}(x) > 0$, then $x = 0$ is unstable.

In the case that $\dot{V}(x) \leq 0$, $V(x)$ is called a Liapunov function. Also if $\dot{V}(x) \equiv 0$, $x \in \Omega$, then the origin is a center, i.e., all solutions lying in Ω are periodic solutions, and their equations in the phase plane are given by the family $V(x) = C$.

E: THE POINCARÉ-BENDIXON THEOREM

The Poincaré-Bendixon theorem is one of the best-known theorems for showing the existence of stable limit cycles of an autonomous system of differential equations in two dimensions. As usual, existence and uniqueness of solutions to initial value problems are assumed. In this case we also assume solutions are extendible (i.e., they exist for all time).

The theorem arises as a consequence of the uniqueness of solutions, which implies that in the phase plane, solutions may not intersect with themselves unless they are periodic or connect with equilibria.

THEOREM (POINCARÉ-BENDIXON) *Let* Γ *be a solution of system* (A.22) *(in two dimensions) and from some time on, let the orbit of* Γ *lie in a closed and bounded set in the plane. Then if the* ω*-limit set of* Γ *does not contain any equilibria, either* Γ *is a periodic orbit of* Γ *approaches a periodic orbit asymptotically.*

The ω-limit set of Γ is the set of all limit points of sequences of the type $(\phi_1(t_n), \phi_2(t_n))$, where such points are on Γ and $\lim_{n\to\infty} t_n = \infty$.

A consequence of the preceding theorem is the following result (also sometimes known as the Poincaré-Bendixon theorem).

Let D *be the annular region lying between the two closed curves* C_1 *and* C_2 *(*C_1 *inside* C_2*) and suppose there are no equilibria in* D.

Suppose, further, that all solutions of system (A.22) *(in two dimensions) which touch* C_2 *cross* C_2 *inward, and all solutions which touch* C_1 *cross* C_1 *outward. Then there is an outermost limit cycle in* D, *asymptotically stable from the outside and an innermost limit cycle in* D, *asymptotically stable from the inside.*

See Fig. A.1 for an illustration of this theorem. An analysis and proof of the Poincaré-Bendixon theorem may be found in Chap. 15 of Coddington and Levinson (1955).

F: THE HOPF-BIFURCATION THEOREM

In this appendix the theorem is discussed for two-dimensional autonomous systems.

Consider the system

$$x' = F(x, \varepsilon) \tag{A.25}$$

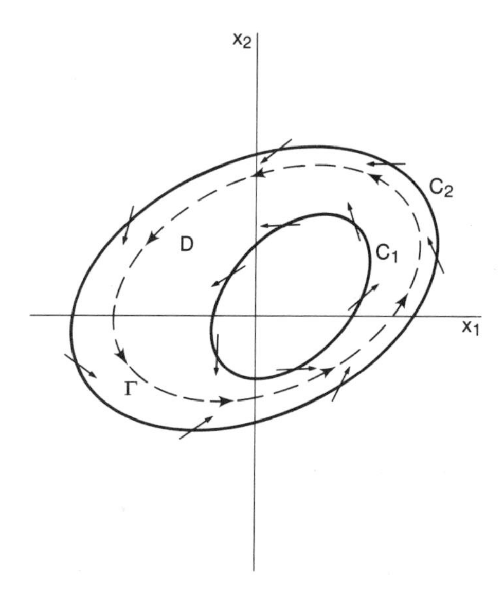

Fig. A.1 All solutions cross inward into D, which is free of equilibria. There are one or more limit cycles Γ stable on at least one side.

where ε is a scalar parameter. Suppose that for sufficiently small ε, there exists a two-dimensional function $a(\varepsilon)$ such that $F(a(\varepsilon),\varepsilon) = 0$ and hence $x = a(\varepsilon)$ is an equilibrium for small ε, depending on ε.

The theorem deals with the circumstance that this equilibrium changes stability at some critical value of ε, which without loss of generality, we take to be $\varepsilon = 0$. Under certain circumstances there will appear "small-amplitude" periodic solutions which may also be limit cycles. To describe these circumstances, we introduce the following notations.

Let $A(\varepsilon)$ be the matrix given by $A(\varepsilon) = F_x(a(\varepsilon),\varepsilon)$. Further, let $A(0) = (\alpha_{ij})$, $A_\varepsilon(0) = \delta_{ij}$ and $A_{\varepsilon\varepsilon}(0) = (\theta_{ij})$. We will also need notations for certain higher derivatives of F. Let $\gamma_{pq}^{(i)} = \partial^2 F_i(a(0),0)/\partial x_1^p \partial x_2^q$, $p + q = 2$, and $\kappa_{pq}^{(i)} = \partial^3 F_i(a(0),0)/\partial x_1^p \partial x_2^q$, $p + q = 3$. Finally, let $\omega = \det |A(0)|$.

The relevant part of the Hopf theorem can be stated as follows.

Let $F(x,\varepsilon)$ *be analytic in* x *and* ε. *Let* $\alpha_{11} + \alpha_{22} = 0$, $\omega > 0$, $\delta_{11} + \delta_{22} \neq 0$. *Then there exists a bifurcation of nontrivial periodic solutions emanating from* $a(\varepsilon)$ *for either* (1) $\varepsilon > 0$ *or* (2) $\varepsilon < 0$ *or* (3) $\varepsilon = 0$. *Further, in cases* (1) *and* (2) *the periodic solutions have the stability opposite to that of the equilibrium on the same side of* $\varepsilon = 0$.

In Freedman (1977) a formula giving a sufficient condition for (3) not to occur was given. Let

$$\begin{aligned} J = \frac{\Pi}{4\omega^5} \Bigg[\Big(& -\alpha_{12}\alpha_{21}\kappa_{30}^{(1)} + 2\alpha_{11}\alpha_{21}\kappa_{21}^{(1)} - \alpha_{21}^2\kappa_{12}^{(1)} - \alpha_{12}\alpha_{21}\kappa_{21}^{(2)} \\ & + 2\alpha_{11}\alpha_{21}\kappa_{12}^{(2)} + \alpha_{21}^2\kappa_{03}^{(2)} \Big)\omega^2 - \alpha_{11}\alpha_{12}\alpha_{21}\gamma_{20}^{(1)^2} \\ & + \alpha_{21}(2\alpha_{11}^2 - \alpha_{12}\alpha_{21})\gamma_{20}^{(1)}\gamma_{11}^{(1)} + \alpha_{11}\alpha_{21}^2\gamma_{20}^{(1)}\gamma_{02}^{(1)} \\ & - \alpha_{12}^2\alpha_{21}\gamma_{20}^{(1)}\gamma_{20}^{(2)} + \alpha_{11}\alpha_{12}\alpha_{21}\gamma_{20}^{(1)}\gamma_{20}^{(2)} + 2\alpha_{11}\alpha_{21}^2\gamma_{11}^{(1)^2} \end{aligned} \qquad (A.26)$$

$$+ \alpha_{21}^3\gamma_{11}^{(1)}\gamma_{02}^{(1)} + \alpha_{11}\alpha_{12}\alpha_{21}\gamma_{11}^{(1)}\gamma_{20}^{(2)} + \alpha_{11}\alpha_{21}^2\gamma_{11}^{(1)}\gamma_{02}^{(2)}$$

$$+ \alpha_{11}\alpha_{21}^2\gamma_{02}^{(1)}\gamma_{11}^{(2)} + \alpha_{21}^3\gamma_{02}^{(1)}\gamma_{02}^{(2)} - \alpha_{12}^2\alpha_{21}\gamma_{20}^{(2)}\gamma_{11}^{(2)}$$

$$+ \alpha_{11}\alpha_{12}\alpha_{21}\gamma_{20}^{(2)}\gamma_{02}^{(2)} + 2\alpha_{11}\alpha_{12}\alpha_{21}{\gamma_{11}^{(2)}}^2$$

$$- \alpha_{21}(2\alpha_{11}^2 - \alpha_{12}\alpha_{21})\gamma_{11}^{(2)}\gamma_{02}^{(2)} - \alpha_{11}\alpha_{21}^2{\gamma_{02}^{(2)}}^2 \Big]$$

Then if $J \neq 0$, case (3) will not occur. Further, it can be shown that if $J > 0$, the periodic solutions are unstable, whereas if $J < 0$, they are asymptotically stable; if $J(\delta_{11} + \delta_{22}) < 0$, then case (2) occurs and if $J(\delta_{11} + \delta_{22}) < 0$, case (1) occurs.

For an excellent account of the Hopf theorem, see Marsden and McCracken (1976).

REFERENCES

ABDELKADER, M. A. (1974): Exact solutions of Lotka-Volterra equations, *Math. Biosci.* 20:293-297.

ABELE, L. G. (1976): Comparative species richness in fluctuating and constant environments: coral-associated decapod crustaceans, *Science* 192:461-463.

ABLES, J. R., and M. SHEPARD (1974): Responses and competition of the parasitoids *Spalangia endius* and *Muscidifurax raptor* (Hymenoptera: Pteromalidae) at different densities of house fly pupae, *Can. Ent.* 106:825-830.

ABRAMS, P. A. (1975): Limiting similarity and the form of the competition coefficient, *Theor. Pop. Biol.* 8:356-375.

______(1976): Niche overlap and environmental variability, *Math. Biosci.* 28:357-372.

ABROSOV, N. S., N. S. PECHURLSIN, and A. V. FURYAEVA (1977): Analysis of food competition of heterotrophic organisms, *Sov. J. Ecol.* 8:505-510.

ADDICOTT, J. E. (1974): Predation and prey community structure: an experimental study of the effect of mosquito larvae on the protozoan communities of pitcher plants, *Ecology* 55:475-492.

AKRE, B. G., and D. M. JOHNSON (1979): Switching and sigmoid functional response curves by damselfly naiads with alternative prey available, *J. Anim. Ecol.* 48:703-720.

ALBRECHT, F., H. GATZKE, A. HADDAD, and N. WAX (1974): The dynamics of two interacting populations, *J. Math. Anal. Appl.* 46:658-670.

______(1976): On the control of certain interacting populations, *J. Math. Anal. Appl.* 53:578-603.

ALBRECHT, F., H. GATZKE, and N. WAX (1973): Stable limit cycles in prey-predator populations, *Science* 181:1073-1074.

ALEKSEEV, V.V., and V. A. SVETLOSANOV (1974): Estimation of the life span of the predator-victim system under random migration of victims, *Sov. J. Ecol.* 5:74-77.

ALLEN, J. C. (1975): Mathematical models of species interactions in time and space, *Amer. Nat.* 109:319-342.

ALLEN, P. M. (1975): Darwinian evolution and a predator-prey ecology, *Bull. Math. Biol.* 37:389-405.

______(1976): Evolution, population dynamics and stability, *Proc. Nat. Acad. Sci.*, USA 73:665-668.

ANDERSEN, F. S. (1960): Competition in populations of one age group, *Biometrics* 16:19-27.

ANDERSON, R. F. V. (1977): Ethological isolation and competition of allospecies in secondary contact, *Amer. Nat.* 111:939-949.

______(1978): Stability of displacement clines arising in allospecies competition, *J. Math. Biol.* 6:131-144.

______(preprint): Evolution of ethological isolation mechanisms.

ANDERSON, R. M., and R. M. MAY (1978): Regulation and stability of host-parasite population interactions I. Regulatory processes, *J. Anim. Ecol.* 47:219-247.

ANDREWARTHA, H. G., and L. C. BIRCH (1953): The Lotka-Volterra theory of interspecific competition, *Austr. J. Zool.* 1:174-177.

ANDRIETTI, F. (1978): Interactions among biological systems: an analysis of asymptotic stability, *Bul. Math. Biol.* 40:839-851.

ANDRONOV, A. A., E. A. LEONTOVICH, I. I. GORDON, and A. G. MAIER (1973): *Qualitative Theory of Second Order Dynamic Systems*, Wiley, New York.

ANTONELLI, P. L., and B. H. VOORHEES (preprint): Geometry of Volterra systems. I: Two species predator-prey models.

ANTONOVICS, J., and H. FORD (1972): Criteria for the validation or invalidation of the competitive exclusion principle, *Nature* 237:406-408.

ARDITI, R., J. M. ABILLON, and J. VIEIRA DA SILVA (1977): The effect of a time-delay in a predator-prey model, *Math. Biosci.* 33:107-120.

______(1978): A predator-prey model with satiation and intraspecific competition, *Ecol. Model* 5:173-191.

ARMSTRONG, R. A. (1976): The effects of predator functional response and prey productivity on predator-prey stability: a graphical approach, *Ecology* 57:609-612.

______(1976): Fugitive species: experiments with fungi and some theoretical considerations, *Ecology* 57:953-963.

______(1977): Weighting factors and scale effects in the calculation of competition coefficients, *Amer. Nat.* 111:810-812.

______(1979): Prey species replacement along a gradient of nutrient enrichment: a graphical approach, *Ecology* 60:76-84.

ARMSTRONG, R. A. and R. P. MCGEHEE (1976a): Coexistence of two competitors on one resource, *J. Theor. Biol.* 56:499-502.

______(1976b): Coexistence of species competing for shared resources, *Theor. Pop. Biol.* 9:317-328.

ARON, J. L. (1979): Harvesting a protected population in an uncertain environment, *Math. Biosci.* 47:197-205.

ARONSON, D. G. (1980): Density dependent interaction--diffusion systems, in *Proceedings of the Advanced Seminar on Dynamics and Modelling of Reactive Systems,* Academic Press, New York.

ASMUSSEN, M. A., and M. W. FELDMAN (1977): Density dependent selection. 1: A stable feasible equilibrium may not be attainable, *J. Theor. Biol.* 64:603-618.

ASSIMACOPOULOS, D., and F. J. EVANS (1979): A qualitiative method for analysis of predator-prey systems under enrichment, *J. Theor. Biol.* 80:467-484.

AUSLANDER, D. M., J. GUCKENHEIMER, and G. F. OSTER (1978): Random evolutionarily stable strategies, *Theor. Pop. Biol.* 13:276-293.

AUSLANDER, D. M., G. F. OSTER, and C. B. HUFFAKER (1974): Dynamics of interacting populations, *J. Franklin Inst.* 297:345-375.

AUSTIN, M. P., and B. G. COOK (1974): Ecosystem stability: a result from an abstract simulation, *J. Theor. Biol.* 45:435-458.

AYALA, F. (1971): Competition between species: frequency dependence, *Science* 171:820-824.

BARCLAY, H. (1975): Population strategies and random environments, *Can. J. Zool.* 53:160-165.

BARCLAY, H., and P. VAN DEN DRIESSCHE (1975): Time lags in ecological systems, *J. Theor. Biol.* 51:347-356.

______(1978): Deterministic population models and stability, *SIAM Review* 20:389-393.

BARTLETT, M. S. (1957): On theoretical models for competitive and predatory biological systems, *Biometrika* 44:27-42.

BARTLETT, M. S., J. C. GOWER, and P. H. LESLIE (1960): A comparison of theoretical and empirical results for some stochastic population models, *Biometrika* 47:1-11.

BARTOZYNSKI, R. (1977): On chances of survival under predation, *Math. Biosci.* 33:135-144.

BARTOZYNSKI, R., and W. BÜHLER (1978): Survival in hostile environments, *Math. Biosci.* 38:293-301.

BAZIN, M. J., V. RAPA, and P. T. SAUNDERS (1974): The integration of theory and experiment in the study of predator-prey dynamics, in M. B. Usher and M. H. Williamson, eds., *Ecological Stability*, Chapman and Hall, London, pp. 159-164.

BAZIN, M. J., and P. T. SAUNDERS (1978): Determination of critical variables in a microbial predator-prey system by catastrophe theory, *Nature* 275:52-54.

BAZYHIN, A. D. (1974): Volterra's system and the Michaelic-Menton equation, in V. A. Ratner, ed., *Problems in Mathematical Genetics,* USSR Academy of Science, (Russian).

BEAVER, R. A. (1974): Intraspecific competition among bark beetle larvae (Coleoptera: Scolytidae), *J. Anim. Ecol.* 43:455-467.

BECK, K. (preprint): A mathematical model of T cell effects in the humoral immune response.

BECKER, N. G. (1973): Interactions between species: some comparisons between deterministic and stochastic models, *Rocky Mt. J. Math.* 3:53-68.

BECUS, G. A. (1979a): Stochastic prey-predator relationships: a random differential equation approach, *Bul. Math. Biol.* 41:91-100.

______(1979b): Stochastic prey-predator relationships: a random evolution approach, *Bul. Math. Biol.* 41:543-554.

BEDDINGTON, J. R. (1974): Age distribution and the stability of simple discrete time population models, *J. Theor. Biol.* 47:65-74.

______(1975): Mutual interference between parasites or predators and its effect on searching efficiency, *J. Anim. Ecol.* 44:331-340.

BEDDINGTON, J. R., and C. A. FREE (1976): Age structure effects in predator-prey interactions, *Theor. Pop. Biol.* 9:15-24.

BEDDINGTON, J. R., C. A. FREE, and J. H. LAWTON (1976): Concepts of stability and resilience in predator-prey models, *J. Anim. Ecol.* 45:791-816.

______(1978): Characteristics of successful natural enemies in models of biological control of insect pests, *Nature* 273:513-519.

BEDDINGTON, J. R., and P. S. HAMMOND (1977): On the dynamics of host-parasite-hyperparasite interactions, *J. Anim. Ecol.* 46:811-821.

BEDDINGTON, J. R., M. P. HASSELL, and J. H. LAWTON (1976): The components of arthropod predation II. The predator rate of increase, *J. Anim. Ecol.* 45:165-185.

BEDDINGTON, J. R., and R. M. MAY (1975): Time delays are not necessarily destabilizing, *Math. Biosci.* 27:109-117.

BEIRNE, B. P. (1975): Biological control attempts by introductions against pest insects in the field in Canada, *Can. Ent.* 107: 225-236.

BELL, G. I. (1970): Mathematical model of clonal selection and antibody production, *J. Theor. Biol.* 29:191-232.

______(1971a): Mathematical model of clonal selection and antibody production II, *J. Theor. Biol.* 33:339-378.

______(1971b): Mathematical model of clonal selection and antibody production III. The cellular basis of immunological paralysis, *J. Theor. Biol.* 33:379-398.

______(1973): Predator-prey equations simulating an immune response, *Math. Biosci.* 16:291-314.

______(1974): Model for the binding of multivalent antigen to cells, *Nature* 248:430-431.

BELLMAN, R. (1953): *Stability Theory of Differential Equations*, McGraw-Hill, New York.

______(1966): *Perturbation Techniques in Mathematics, Physics, and Engineering*, Holt, Rinehart and Winston, New York.

BENKE, A. C. (1978): Interactions among coexisting predators--a field experiment with dragonfly larvae, *J. Anim. Ecol.* 47: 335-350.

BEYER, W. A., D. R. HARRIS, and R. J. RYAN (1979): A stochastic model of the Isle Royale biome, *Rocky Mountain J. Math.* 9:3-18.

BHARGAVA, S. C., and R. P. SAXENA (1977): Stable periodic solutions of the reactive-diffusive Volterra system of equations, *J. Theor. Biol.* 67:399-407.

BILLARD, L. (1977): On the Lotka-Volterra predator-prey models, *J. Appl. Prob.* 14:375-381.

BIGGER, M. (1973): An investigation by Fourier analysis into the interaction between coffee leaf-miners and their larval parasites, *J. Anim. Ecol.* 42:417-434.

______(1976): Oscillations of tropical insect populations, *Nature* 259:207-209.

BIRCH, L. C. (1957): The meanings of competition, *Amer. Nat.* 91: 5-18.

BIRKHOFF, G., and G. C. ROTA (1969): *Ordinary Differential Equations*, Blaisdell, Waltham, Mass.

BIRKLAND, C. (1974): Interactions between a sea pen and seven of its predators, *Ecol. Monogr.* 44:211-232.

BIRLEY, M. H. (1979): The theoretical control of seasonal pests--a single species model, *Math. Biosci.* 43:141-157.

BISHOP, J. A., and L. M. COOK (1975): Moths, melanism and clean air, *Sci. Amer.* 32:90-99.

BISWAS, S. N., K. C. GUPTA, and B. B. KARMAKER (1977): Brillouin-Wigner perturbation solution of Volterra's prey-predator system, *J. Theor. Biol.* 64:253-260.

BOBISUD, L. E. (1976): Canibalism as an evolutionary strategy, *Bull. Math. Biol.* 38:359-368.

BOBISUD, L. E. and W. L. VOXMAN (1979): Predator response to variation of prey density in a patchy environment: a model, *Amer. Nat.* 114:63-75.

BODENHEIMER, F. S., and M. SCHIFFER (1952): Mathematical studies in animal populations. I: A mathematical study of insect parasitism, *Acta Biotheor.* 10:23-56.

BODJADZIEV, G. (1978): The Krylov-Bogoliubov-Mitropolskii method applied to models of population dynamics, *Bul. Math. Biol.* 40:335-345.

BODJADZIEV, G., and S. CHAN (1979): Asymptotic solutions of differential equations with time delay in population dynamics, *Bul. Math. Biol.* 41:325-342.

BOTKIN, D. B., and M. J. SOBEL (1975): Stability in time-varying ecosystems, *Amer. Nat.* 109:625-646.

BOVBJERG, R. V. (1970): Ecological isolation and competitive exclusion in two crayfish (*Orconectes virilis* and *Orconectes immunis*), *Ecology* 51:225-236.

BOUNDS, J. M. and J. M. CUSHING (1975): On the behaviour of solutions of predator-prey equations with hereditary terms, *Math. Biosci.* 26:41-54.

BRADDOCK, R. D., and P. VAN DER DRIESSCHE (1976): On the stability of differential-difference equations, *J. Austral. Math. Soc.* 19:358-370.

BRAUER, F. (1972): A nonlinear predator-prey problem, in L. Weiss, ed., *Ordinary Differential Equations*, Academic Press, New York, 1972:371-377.

______(1974): On the populations of competing species, *Math. Biosci.* 19:299-306.

______(1975): On a nonlinear integral equation for population growth problems, *SIAM J. Math. Anal.* 6:312-317.

______(1976a): Constant rate harvesting of populations governed by Volterra integral equations, *J. Math. Anal. Appl.* 56:18-27.

______(1976b): De-stabilization of predator-prey systems under enrichment, *Int. J. Control* 23:541-552.

______(1979a): Harvesting strategies for population systems, *Rocky Mountain J. Math.* 9:19-26.

______(1979b): Boundedness of solutions of predator-prey systems, *Theor. Pop. Biol.* 15:268-273.

______(preprint,a): Periodic solutions of some ecological models.

______(preprint,b): Stability of some population models with delay.

BRAUER, F., and D. A. SANCHEZ (1975a): Constant rate population harvesting: equilibrium and stability, *Theor. Pop. Biol.* 8:12-30.

______(1975b): Some models for population growth with harvesting, in H. A. Antosiewicz, ed., *International Conference on Differential Equations*, Academic Press, pp. 53-64.

BRAUER, F., and A. C. SOUDAK (1979a): Stability regions and transition phenomena for harvested predator-prey systems, *J. Math. Biol.* 7:319-337.

______(1979b): Stability regions in predator-prey systems with constant rate harvesting, *J. Math. Biol.* 8:55-71.

BRAUER, F., A. C. SOUDAK, and H. S. JAROSCH (1976): Stabilization and de-stabilization of predator-prey systems under harvesting and nutrient enrichment, *Int. J. Control* 23:553-573.

BRENCHLEY, G. A. (1979): Community matrix models: inconsistent results using Vandermeer's data, *Amer. Nat.* 113:456-459.

BROCKELMAN, W. Y., and R. M FAGEN (1972): On modeling density-independent population change, *Ecology* 53:944-948.

BRUNI, C., A. GERMANI, G. KOCH, and B. STROM (1976): Derivation of antibody distribution from experimental binding data, *J. Theor. Biol.* 6:143-170.

BRUNI, C., M. A. GIOVENCO, G. KOCH, and R. STROM (1975): A dynamical model of humoral immune response, *Math. Biosci.* 27:191-211.

BULMER, M. G. (1976): The theory of prey-predator oscillations, *Theor. Pop. Biol.* 9:137-150.

BURNETT, T. (1979): An acarine predator-prey population infesting roses, *Res. Pop. Ecol.* 20:227-234.

BUSS, L. W., and J. B. C. JACKSON (1979): Competitive networks: nontransitive competitive relationships in cryptic coral reef environments, *Amer. Nat.* 113:223-234.

BUTLER, G. J. (preprint): Coexistence in predator-prey systems.

BUTLER, G. J., and H. I. FREEDMAN (1972): Further critical cases of the implicit function theorem, *Aequationes Math.* 8:203-211.

BUTLER, G. J., and P. WALTMAN (preprint): Existence of a periodic solution for a two predator-one prey ecosystem modelled on a chemostat.

CAMERON, A. W. (1964): Competitive exclusion between the rodent genera Microtus and Clethrionamys, *Evolution* 18:630-634.

CAMERON, R. A. D., and M. A. CARTER (1979): Intra- and interspecific effects of population density on growth and activity in some helicoid land snails (Gastropoda: Pulmonata), *J. Anim. Ecol.* 48:237-246.

CANALE, R. P. (1970): An analysis of models describing predator-prey interaction, *Biotech. Bioeng.* 12:353-378.

CAPERON, J. (1969): Time lag in population growth response of *Isochrysis galbana* to a variable nitrate environment, *Ecology* 50:188-192.

CARBYN, L. N. (1974): Wolf population fluctuations in Jasper National Park, Alberta, Canada, *Biol. Conserv.* 6:94-101.

CASE, T. J. (1980): MacArthur's minimization principle: a footnote, *Amer. Nat.* 115:133-138.

CASE, T. J., and R. G. CASTEN (1979): Global stability and multiple domains of attraction in ecological systems, *Amer. Nat.* 113:705-714.

CASE, T. J., M. E. GILPIN, and J. M. DIAMOND (1979): Overexploitation, interference competition, and excess density compensation in insular faunas, *Amer. Nat.* 113:843-854.

CASE, T. J., and R. K. WASHINO (1979): Flatworm control of mosquito larvae in rice fields, *Science* 206:1412-1414.

CASWELL, H. (1972): A simulation study of a time lag population model, *J. Theor. Biol.* 34:419-439.

______(1978): Predator mediated coexistence: a non-equilibrium model, *Amer. Nat.* 112:127-154.

CESARI, L. (1971): *Asymptotic Behavior and Stability Problems in Ordinary Differential Equations,* Springer-Verlag, New York.

CHAN, W-Y., AND K. F. WALLES (1978): Multiple time series modelling: another look at the mink-muskrat interaction, *Appl. Statist.* 27:168-175.

CHENG, K. S., S. B. HSU, and S. S. LIN (preprint): Some results on global stability of a predator-prey system.

CHESSON, P. (1978): Predator-prey theory and variability, *Ann. Rev. Ecol. Syst.* 9:323-347.

CHEWNING, W. C. (1975): Migratory effects in predator-prey models, *Math. Biosci.* 23:253-262.

CHOW, P. L., and W. C. TAM (1976): Periodic and travelling wave solutions to Volterra-Lotka equations with diffusion, *Bull. Math. Biol.* 38:643-658.

CHUA, T. H. (1979): A comparative study of the searching efficiencies of a parasite and a hyperparasite, *Res. Pop. Ecol.* 20:179-187.

CLARK, J. P. (1971): The second derivative and population modeling, *Ecology* 52:605-613.

CLARKE, B. (1973): Mutation and population size, *Heredity* 31:367-379.

CLERC, R., C. HARTMAN, and C. MIRA (1977): Transition "order to chaos" in a predator-prey model in the form of a recurrence, *Informatica* 77:1-4.

CODDINGTON, E. A., and N. LEVINSON (1955): *Theory of Ordinary Differential Equations,* McGraw-Hill, New York.

CODY, M. L. (1974): *Competition and the Structure of Bird Communities,* Princeton University Press.

COFFMAN, C. V., and B. D. COLEMAN (1979): On the growth of populations with narrow spread in reproductive age: III. Periodic variations in the environment, *J. Math. Biol.* 7:281-301.

COHEN, D. S., E. COUTSIAS, and J. C. NEU (1979): Stable oscillations in single species growth models with hereditary effects, *Math. Biosci.* 44:255-268.

COHEN, D. S., P. S. HAGEN, and H. C. SIMPSON (1979): Spacial structures in predator-prey communities with hereditary effects and diffusion, *Math. Biosci.* 44:167-177.

COHEN, D. S., and S. ROSENBLAT (1979): Multi-species interactions with hereditary effects and spacial diffusion, *J. Math. Biol.* 7:231-241.

COHEN, J. E. (1970): A Markov contingency-table model for replicated Lotka-Volterra systems near equilibrium, *Amer. Nat.* 104:547-560.

______(1977a): Ratio of prey to predators in community food webs, *Nature* 270:165-167.

______(1977b): Food webs and the dimensionality of trophic niche space, *Proc. Nat. Acad. Sci.* USA 74:4533-4536.

______(1978): *Food Webs and Niche Spaces,* Princeton University Press, Princeton.

______(1979): Graph theoretic model of food webs, *Rocky Mountain J. Math.* 9:29-30.

COHEN, S. (1971): Receptor theory and antibody formation, in *Cellular Interations in the Immune Response,* Karger, Basel, pp. 132-139.

COLE, L. C. (1960): Competitive exclusion, *Science* 132:348-349.

COLEMAN, B. D., Y-H. HSIEH, and G. P. KNOWLES (1979): On the optimal choice of r for a population in a periodic environment, *Math. Biosci.* 46:71-85.

COLWELL, R. K. (1973): Competition and coexistence in a simple tropical community, *Amer. Nat.* 107:737-760.

COLWELL, R. K., and D. J. FUTUYMA (1971): On the measurement of niche breadth and overlap, *Ecology* 52:567-576.

COMINS, H. N., and D. W. E. BLATT (1974): Prey-predator models in spatially heterogeneous environments, *J. Theor. Biol.* 48:75-83.

COMINS, H. N., and M. P. HASSELL (1976): Predation in multi-prey communities, *J. Theor. Biol.* 62:93-114.

CONRAD, M. (1972): Stability of foodwebs and its relation to species diversity, *J. Theor. Biol.* 34:325-335.

CONWAY, E. D., and J. A. SMALLER (1977): Diffusion and the predator-prey interaction, *SIAM J. Appl. Math.* 33:673-686.

______(preprint): Diffusion and the classical ecological interactions: asymptotics.

COOKE, D., and J. A. LEON (1976): Stability of population growth determined by 2 × 2 Leslie matrix with density-dependent elements, *Biometrics* 32:435-442.

COOKE, K. L., and J. A. YORKE (1972): Equations modelling population growth, economic growth, and gonorrhea epidemiology, in L. Weiss, ed., *Ordinary Differential Equations,* Academic Press:35-53.

______(1973): Some equations modelling growth processes and gonorrhea epidemics, *Math. Biosci.* 16:75-101.

CORNELL, H. (1976): Search strategies and the adaptive significance of switching in some general predators, *Amer. Nat.* 110:317-320.

CORNELL, H., and D. PIMENTEL (1978): Switching in the parasitoid Nosonia Vitripennis and its effects on host competition, *Ecology* 59:297-308.

COSTE, J., J. PEYRAUD, and P. COULLET (1978): Does complexity favor the existence of persistent ecosystems?, *J. Theor. Biol.* 73:359-362.

______(1979): Asymptotic behaviors in the dynamics of competing species, *SIAM J. Appl. Math.* 36:516-543.

COSTE, J., J. PEYRAUD, P. COULLET, and A. CHENCINER (1978): About the theory of competing species, *Theor. Pop. Biol.* 14:165-184.

COULMAN, G. A., S. R. REICE, and R. L. TUMMALA (1971): Population modelling: a systems approach, *Science* 175:518-521.

COUTLEE, E. L., and R. I. JENNRICH (1968): The relevance of logarithmic models for population interaction, *Amer. Nat.* 102:307-321.

CRAMER, N. F., and R. M. MAY (1972): Interspecific competition, predation and species diversity: a comment, *J. Theor. Biol.* 34:289-293.

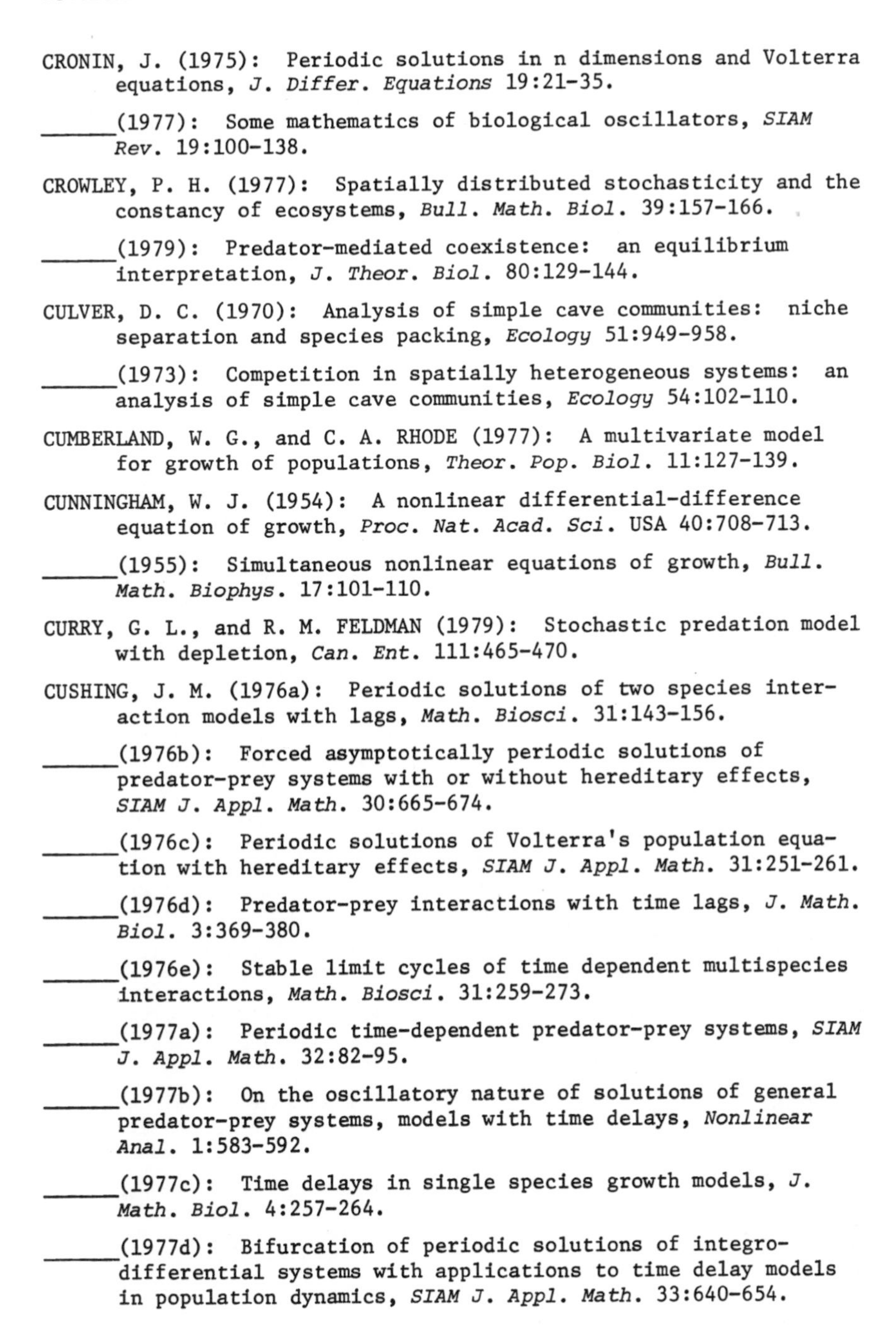

CRONIN, J. (1975): Periodic solutions in n dimensions and Volterra equations, *J. Differ. Equations* 19:21-35.

______(1977): Some mathematics of biological oscillators, *SIAM Rev.* 19:100-138.

CROWLEY, P. H. (1977): Spatially distributed stochasticity and the constancy of ecosystems, *Bull. Math. Biol.* 39:157-166.

______(1979): Predator-mediated coexistence: an equilibrium interpretation, *J. Theor. Biol.* 80:129-144.

CULVER, D. C. (1970): Analysis of simple cave communities: niche separation and species packing, *Ecology* 51:949-958.

______(1973): Competition in spatially heterogeneous systems: an analysis of simple cave communities, *Ecology* 54:102-110.

CUMBERLAND, W. G., and C. A. RHODE (1977): A multivariate model for growth of populations, *Theor. Pop. Biol.* 11:127-139.

CUNNINGHAM, W. J. (1954): A nonlinear differential-difference equation of growth, *Proc. Nat. Acad. Sci.* USA 40:708-713.

______(1955): Simultaneous nonlinear equations of growth, *Bull. Math. Biophys.* 17:101-110.

CURRY, G. L., and R. M. FELDMAN (1979): Stochastic predation model with depletion, *Can. Ent.* 111:465-470.

CUSHING, J. M. (1976a): Periodic solutions of two species interaction models with lags, *Math. Biosci.* 31:143-156.

______(1976b): Forced asymptotically periodic solutions of predator-prey systems with or without hereditary effects, *SIAM J. Appl. Math.* 30:665-674.

______(1976c): Periodic solutions of Volterra's population equation with hereditary effects, *SIAM J. Appl. Math.* 31:251-261.

______(1976d): Predator-prey interactions with time lags, *J. Math. Biol.* 3:369-380.

______(1976e): Stable limit cycles of time dependent multispecies interactions, *Math. Biosci.* 31:259-273.

______(1977a): Periodic time-dependent predator-prey systems, *SIAM J. Appl. Math.* 32:82-95.

______(1977b): On the oscillatory nature of solutions of general predator-prey systems, models with time delays, *Nonlinear Anal.* 1:583-592.

______(1977c): Time delays in single species growth models, *J. Math. Biol.* 4:257-264.

______(1977d): Bifurcation of periodic solutions of integro-differential systems with applications to time delay models in population dynamics, *SIAM J. Appl. Math.* 33:640-654.

______(1978): Bifurcation of periodic oscillations due to delays in single species growth models, *J. Math. Biol.* 6:145-161.

______(1979): Stability and instability in predator-prey models with growth rate response delays, *Rocky Mountain J. Math.* 9:43-50.

DARLINGTON, P. J., Jr. (1972): Competition, competitive repulsion, and coexistence, *Proc. Nat. Acad. Sci.* USA 69:3151-3155.

DEAKIN, M. A. B. (1975): The steady states of ecosystems, *Math. Biosci.* 24:319-331.

DEAN, J. M., and R. E. RICKLEFS (1979): Do parasites of Lepidoptera larvae compete for hosts? No!, *Amer. Nat.* 113: 302-306.

DE ANGELIS, D. L. (1975a): Global asymptotic stability criteria for models of density-dependent population growth, *J. Theor. Biol.* 50:35-43.

______(1975b): Stability and connectance in food web models, *Ecology* 56:238-243.

______(1975c): Estimates of predator-prey limit cycles, *Bull. Math. Biol.* 37:291-299.

______(1975d): Application of stochastic models to a wildlife population, *Math. Biosci.* 31:227-236.

DE ANGELIS, D. L., and R. A. GOLDSTEIN (1978): Criteria that forbid a large, nonlinear food-web model from having more than one equilibrium point, *Math. Biosci.* 41:81-90.

DE ANGELIS, D. L., R. A. GOLDSTEIN, and R. V. O'NEILL (1975): A model for trophic interactions, *Ecology* 56:881-892.

DE ANGELIS, D. L., C. C. TRAVIS, and W. M. POST (1979): Persistence and stability of seed-dispersed species in a patchy environment, *Theor. Pop. Biol.* 16:107-125.

DE BACH, P., and H. S. SMITH (1941): Are population oscillations inherent in the host parasite relation?, *Ecology* 22:363-369.

DE BENEDICTIS, P. A. (1974): Interspecific competition between tadpoles of *Rana pipiens* and *Rana sylvatica*: an experimental field study, *Ecol. Monogr.* 44:129-151.

DEISTLER, M., and G. FEICHTINGER (1974): The linear model formulation of a multitype branching process applied to population dynamics, *J. Amer. Stat. Assoc.* 69:662-664.

DE JONG, G. (1976): A model of competition for food. I: Frequency-dependent viabilities, *Amer. Nat.* 110:1013-1027.

DEKKER, H. (1975): A simple mathematical model of rodent population cycles, *J. Math. Biol.* 2:57-67.

DELATTRE, P. (1976): Transformation system with time-dependent characteristic and population theory, *Math. Biosci.* 32:239-274.

DE LISI, C. (1977): Detection and analysis of recognition and selection in the immune response, *Bull. Math. Biol.* 39:705-719.

DE LISI, C., and A. RESCIGNO (1977): Immune surveillance and neoplasia. I: A minimal mathematical model, *Bull. Math. Biol.* 39:201-221.

DE MONTONI, P., and F. ROTHE (1979): Convergence to homogeneous equilibrium state for generalized Volterra-Lotka systems with diffusion, *SIAM J. Appl. Math.* 37:648-663.

DENNIS, B. (1978): Analytical solution to an open-system model of population grwoth, *Math. Biosci.* 40:167-169.

DE VITA, J. (1979): Niche separation and the broken-stick model, *Amer. Nat.* 114:171-178.

DHONT, A. A. (1977): Interspecific competition between great and blue tit, *Nature* 268:521-523.

DIAMOND, P. (1974a): The stability of the interaction between entomophagous parasites and their host, *Math. Biosci.* 19:121-129.

______(1974b): Stochastic stability of a host-parasite model, *Math. Biosci.* 22:339-343.

______(1974c): Area of discovery of an insect parasite, *J. Theor. Biol.* 45:467-471.

______(1976): Domains of stability and resilience for biological populations obeying difference equations, *J. Theor. Biol.* 61:287-306.

DI BLASIO, G., and L. LAMBERTI (1978): An initial-boundary value problem for age-dependent population diffusion, *SIAM J. Appl. Math.* 35:593-615.

DIBROV, B. F., M. A. LIVSHITS, and M. V. VOLKENSTEIN (1977): Mathematical model of immune processes, *J. Theor. Biol.* 65: 609-631.

DIXON, K. R., and G. W. CORNWELL (1970): A mathematical model for predator and prey populations, *Res. Pop. Ecol.* 12:127-136.

DUTT, R. (1976): Application of Hamilton-Jacobi theory to the Lotka-Volterra oscillator, *Bull. Math. Biol.* 38:459-465.

DUTT, R., and P. K. GHOSH (1975): Nonlinear correction to Lotka-Volterra oscillation in a prey-predator system, *Math. Biosci.* 27:9-16.

DUTT, R., P. K. GHOSH, and B. B. KARMAKER (1975): Application of perturbation theory to the nonlinear Volterra-Gause-Witt model for prey-predator interaction, *Bull. Math. Biol.* 37: 139-146.

EHLER, L. E., and R. VAN DEN BOSCH (1974): An analysis of the natural biological control of *Trichoplusia ni* (Lepidoptera: Noctuidae) on cotton in California, *Can. Ent.* 106:1067-1073.

EISEN, M., and J. SCHILLER (1977): Stability analysis of normal and neoplastic growth, *Bull. Math. Biol.* 39:597-605.

ENGSTROM-HEG, V. L. (1970): Predation, competition and environmental variables: some mathematical models, *J. Theor. Biol.* 27:175-195.

EPSTEIN, I. R. (1979): Competitive coexistence of selfreproducing macromolecules, *J. Theor. Biol.* 78:271-298.

EVANS, G. T. (1977): Functional response and stability, *Amer. Nat.* 111:799-802.

EVERSON, P. (1979): The functional response of Phytoseiulus persimilis (Acarina: Phytoseüdae) to various densities of Tetranychus urticae (Acarina: Tetranychidae), *Can. Ent.* 111:7-10.

FELDMAN, M. W., and J. ROUGHGARDEN (1975): A population's stationary distribution and chance of extinction in a stochastic environment with remarks on the theory of species packing, *Theor. Pop. Biol.* 7:197-207.

FELLER, W. (1940): On the logistic law of growth and its empirical verification in biology, *Acta Biotheor.* 5:51-65.

FELSENSTEIN, J. (1979): r- and K-selection in a completely chaotic population model, *Amer. Nat.* 113:499-510.

FENCHEL, T. M., and F. B. CHRISTIANSEN (1977): Selection and interspecific competition, in F. B. Christiansen and T. M. Fenchel, eds. *Measuring Selection in Natural Populations,* Springer-Verlag Lecture Notes in Biomathematics, Vol. 19, New York, pp. 477-498.

FINERTY, J. P. (1979): Cycles in Canadian lynx, *Amer. Nat.* 114: 453-455.

FISHER, M. E., and B. S. GOH (1977): Stability in a class of discrete time models of interacting populations, *J. Math. Biol.* 4:265-274.

FISHER, M. E., B. S. GOH, and T. L. VINCENT (1979): Some stability conditions for discrete-time single species models, *Bul. Math. Biol.* 41:861-875.

FLAHERTY, D. L. (1969): Ecosystem complexity and densities of the Willamette mite, *Eotetranychus willametti* Ewing (Acarina: Tetranychidae), *Ecology* 50:911-916.

FLAHERTY, D. L., and C. B. HUFFAKER (1970a): Biological control of Pacific mites and Willamette mites in San Joaquin Valley vineyards. I: Role of *Metaseuilus occidentalis, Hilgardia* 40:267-308.

______(1970b): Biological control of Pacific mites and Willamette mites in San Joaquin Valley vineyards. II: Influence of dispersion patterns of *Metaseiulus occidentalis*, *Hilgardia* 40:309-330.

FORCE, D. C. (1974): Ecology of insect host-parasitoid communities, *Science* 184:624-632.

FOSTER, C., and A. RAPOPORT (1956): Parasitism and symbiosis in an N-person non-constant-sum continuous game, *Bull. Math. Biophys.* 16:219-231.

FRAME, J. S. (1974): Explicit solutions in two species Volterra systems, *J. Theor. Biol.* 43:73-81.

FREDERICKSON, A. G., J. L. JOST, H. M. TSUCHIYA, and P.-H. HSU (1973): Predator-prey interactions between malthusian populations, *J. Theor. Biol.* 38:487-526.

FREE, C. A., J. R. BEDDINGTON, and J. H. LAWTON (1977): On the inadequacy of simple models of mutual interference for parasitism and predation, *J. Anim. Ecol.* 46:543-554.

FREEDMAN, H. I. (1969): The implicit function theorem in the scalar case, *Can. Math. Bull.* 12:721-732.

______(1975): A perturbed Kolmogorov-type model for the growth problem, *Math. Biosci.* 23:127-149.

______(1976): Graphical stability, enrichment, and pest control by a natural enemy, *Math. Biosci.* 3:207-225.

______(1977): On a bifurcation theorem of Hopf and Friedrichs, *Can. Math. Bull.* 20:95-102.

______(1979): Stability analysis of a predator-prey system with mutual interference and desnity-dependent death rates, *Bull. Math. Biol.* 41:67-78.

FREEDMAN, H. I., and J. A. GATICA (1977): A threshold model simulating humoral immune response to replicating antigens, *Math. Biosci.* 37:113-134.

______(1979): On the boundedness of lymphocytes in deterministic threshold models of humoral immune response, *Rocky Mt. J. Math.* 9:73-81.

FREEDMAN, H. I., and P. WALTMAN (1975a): Perturbation of two dimensional predator-prey equations, *SIAM J. Appl. Math.* 28:1-10.

______(1975b): Periodic solutions of perturbed Lotka-Volterra systems, in H. A. Antosiewicz, ed., *International Conference on Differential Equations*, Academic Press, New York, pp. 312-316.

______(1975c): Perturbation of two dimensional predator-prey equations with an unperturbed critical point, *SIAM J. Appl. Math.* 29:719-733.

______(1977a): Mathematical models of species interaction with dispersal I: Stability of two habitats with and without a predator, *SIAM J. Appl. Math.* 32:631-648.

______(1977b): Mathematical analysis of some three species food chain models, *Math. Biosci.* 33:257-276.

______(1978): Predator influence on the growth of a population with three genotypes, *J. Math. Biol.* 6:367-374.

FRETWELL, S. (1978): Competition for discrete versus continuous resources: tests for predictions from the MacArthur - Levins models, *Amer. Nat.* 112:73-81.

FRISMAN, E. YA. (1980): Difference in densities of individuals in populations with uniform ranges, *Ecol. Model.* 8:345-354.

FUJI, K. (1977): Complexity-stability relationship of two-prey-one-predator species system model: local and global stability, *J. Theor. Biol.* 69:613-623.

FUJITA, K., T. INOUE, and A. TAKAFUJI (1979): Systems analysis of an acarine predator-prey system. I., *Res. Pop. Ecol.* 21:105-119.

FUYTUYMA, D. J. (1973): Community structure and stability in constant environments, *Amer. Nat.* 107:443-446.

GAFFNEY, P. M. (1975): Roots of the niche concept, *Amer. Nat.* 109:490.

GALE, J. S. (1964): Competition between three lines of Drosophila melanogaster, *Heredity* 11:681-699.

GALLOPIN, G. C. (1972): Trophic similarity between species in a food web, *Amer. Midl. Nat.* 87:336-343.

GANDOLFI, A., M. A. GIOVENCO, and R. STROM (1978): Reversible binding of multivalent antigen in the control of B lymphocyte activation, *J. Theor. Biol.* 74:513-521.

GARD, T. C. and T. G. HALLAM (1979): Persistence in food webs. I: Lotka-Volterra food chains, *Bul. Math. Biol.* 41:877-891.

GARD, T. C. and D. KANNAN (1976): On a stochastic differential equation modelling of predator-prey evolution, *J. Appl. Prob.* 13:429-443.

GARFINKEL, D. (1962): Digital computer simulation of ecological systems, *Nature* 194:856-857.

______(1967): Effect on stability of Lotka-Volterra ecological systems of imposing strict territorial limits on populations, *J. Theor. Biol.* 14:325-327.

GARFINKEL, D., and R. SACK (1964): Digital computer simulation of an ecological system, based on a modified mass action law, *Ecology* 45:502-507.

GATICA, J. A., and H. L. SMITH (1977): Fixed point techniques in a cone with applications, *J. Math. Anal. Appl.* 6:58-71.

GATICA, J. A., and P. E. WALTMAN (1976): A singular functional differential equation arising in an immunological model, in W. N. Everitt and B. D. Sleeman, eds., *Ordinary and Partial Differential Equations,* Springer-Verlag Lecture Notes in Mathematics, Vol. 564, New York, pp. 114-124.

______(1978): A threshold model of antigen-antibody dynamics, *Proceedings of the 1978 IEEE Conference on Decision and Control*:746-748.

GATTO, M., and S. RINALDI (1977): Stability analysis of predator-prey models via the Liapunov method, *Bull. Math. Biol.* 39: 339-347.

GAUSE, G. F. (1934): *The Struggle for Existence*, Williams and Wilkins, Baltimore, 1934.

GAUSE, G. F., N. P. SMARAGDOVA, and A. A. WITT (1936): Further studies of interaction between predators and prey, *J. Anim. Ecol.* 5:1-18.

GAUSE, G. F., and A. A. WITT (1935): Behavior of mixed populations and the problem of natural selection, *Amer. Nat.* 69:596-609.

GAZIS, D. C., E. W. MONTROLL and J. E. RYNIKER (1973): Age-specific deterministic model of predator-prey populations: Application to Isle Royale, *IBM J. Res. Develop.* 17:47-53.

GETZ, W. M. (1976): Stochastic equivalents of the linear and Lotka-Volterra systems of equations--a general birth-and-death process formulation, *Math. Biosci.* 29:235-257.

______(1979): On harvesting two competing populations, *J. Opt. Theory Applic.* 28:585-602.

______(1980): The ultimate-sustainable-yield problem in nonlinear age-structural populations, *Math. Biosci.* 48:279-292.

GHABBOUR, S. I. (1972-74): Insecticides and cotton in Egypt, *Biol. Conserv.* 5-6:62-63.

GHENT, A. W. (1960): A study of the group-feeding behavior of larvae of the jack pine sawfly, *Neodiprian pratti banksianae Roh., Behaviour* 16:110-148.

GILL, D. E. (1972): Intrinsic rates of increase, saturation densities, and competitive ability. I: An experiment with Paramecium, *Amer. Nat.* 106:461-471.

______(1974): Intrinsic rate of increase, saturation density, and competitive ability. II: The evolution of competitive ability, *Amer. Nat.* 108:103-116.

______(1978): On selection at high population density, *Ecology* 59:1289-1291.

GILPIN, M. E. (1972): Enriched predator-prey systems: theoretical stability, *Science* 177:902-904.

______(1973): Do hares eat lynx? *Amer. Nat.* 107:727-730.

______(1974a): A Liapunov function for competition communities, *J. Theor. Biol.* 44:35-48.

______(1974b): A model of the predator-prey relationships, *Theor. Pop. Biol.* 5:333-344.

______(1974c): Intraspecific competition between Drosophila larvae in serial transfer systems, *Ecology* 55:1154-1159.

______(1974d): Habitat selection and a Liapunov function for competition communities, in P. Van den Driesshe, ed., *Mathematical Problems in Biology,* Springer-Verlag Lecture Notes in Biomathematics, Vol. 2, New York, pp. 62-65.

______(1975a): Limit cycles in competition communities, *Amer. Nat.* 109:51-60.

______(1975b): *Group Selection in Predator-Prey Communities,* Princeton University Press, Princeton, N. J.

______(1979a): Prudent predation and the character of ecological attractor sets, *Rocky Mountain J. Math.* 9:83-86.

______(1979b): Spiral chaos in a predator-prey model, *Amer. Nat.* 113:306-308.

GILPIN, M. E., and F. J. AYALA (1976): Schoener's model and Drosophila competition, *Theor. Pop. Biol.* 9:12-14.

GILPIN, M. E., T. J. CASE, and F. J. AYALA (1976): θ-Selection, *Math. Biosci.* 32:131-139.

GILPIN, M. E., and J. M. DIAMOND (1976): Calculation of immigration and extinction curves from the species-area-distance relation, *Proc. Nat. Acad. Sci. USA* 73:4130-4134.

GILPIN, M. E., and K. E. JUSTICE (1973): A note on nonlinear competition, *Math. Biosci.* 17:57-63.

GINZBERG, L. R. (1977): Local consideration of polymorphisms for populations coexisting in stable ecosystems, *J. Math. Biol.* 5:33-41.

GLASSER, J. W. (1978): The effect of predation on prey resource utilization, *Ecology* 59:724-732.

______(1979): The role of predation in shaping and maintaining the structure of communities, *Amer. Nat.* 113:631-641.

GLEIT, A. (1978): Optimal harvesting in continuous time with stochastic growth, *Math. Biosci.* 41:111-123.

______(preprint): On optimal mean variance harvesting rules.

GLEN, D. M. (1975): Searching behaviour and prey-density requirements of *Blepharidopterus angulatus* (Fall.) (Heteroptera: Miridae) as a predator of the lime aphid, *Eucallipterus tiliae* (L.) and leafhopper, *Alnetoidea alneti* (Dahlbom), *J. Anim. Ecol.* 44:115-134.

GLEN, D. M., and P. BRAIN (1978): A model of predation on codling moth eggs (Cydia pomonella), *J. Anim. Ecol.* 42:711-724.

GLIDDON, C., and C. STROBECK (1975): Necessary and sufficient conditions for multiple-niche polymorphism in haploids, *Amer. Nat.* 109:233-235.

GOEL, N. S., S. C. MAITRA, and E. W. MONTROLL (1971): On the Volterra and other nonlinear models of interacting populations, *Rev. Mod. Phys.* 43:231-276.

GOH, B. S. (1969-70): Optimal control of a fish resource, *Malayan Sci.* 5:65-70.

______(1973): Optimal control of renewable resources and pest populations, *Proc. Sixth Hawaii Int. Conf. Syst. Sci.*, pp. 26-28.

______(1974): Stability and diversity in ecosystems, *Proc. Seventh Hawaii Int. Conf. Syst. Sci.*, pp. 241-243.

______(1975): Stability, vulnerability and persistence of complex ecosystems, *Ecol. Model.* 1:105-116.

______(1976a): Nonvulnerability of ecosystems in unpredictable environments, *Theor. Pop. Biol.* 10:83-95.

______(1976b): Global stability in two species interactions, *J. Math. Biol.* 3:313-318.

______(1977): Global stability in many species systems, *Amer. Nat.* 111:135-143.

______(1978a): Sector stability of a complex ecosystem model, *Math. Biosci.* 40:157-166.

______(1978b): Robust stability concepts of ecosystem models, in E. Halfen, ed. *Theoretical Systems Ecology*, Academic Press, New York.

______(1978c): Global stability in a class of prey-predator models, *Bul. Math. Biol.* 40:525-533.

______(1978d): The usefulness of optimal control theory to ecological problems, in E. Halfon, ed. *Theoretical Systems Ecology*, Academic Press, New York.

______(1979): Stability in models of mutualism, *Amer. Nat.* 113: 261-275.

GOH, B. S., and T. T. AGNEW (1977): Stability in Gilpin and Ayala's models of competition, *J. Math. Biol.* 4:275-279.

GOH, B. S., and L. S. JENNINGS (to appear): Feasibility and stability in randomly assembled Lotka-Volterra models, *Ecol. Model.*

GOH, B. S., G. LEITMAN, and T. L. VINCENT (1974): Optimal control of a prey-predator system, *Math. Biosci.* 19:263-286.

GOH, B. S., T. L. VINCENT, and D. J. WILSON (1974a): Suboptimal control of prey-predator system by graphical analysis, *Proc. Seventh Hawaii Int. Conf. Syst. Sci.*, pp. 196-199.

______(1974b): A method for formulating suboptimal policies for crudely modelled ecosystems, *Proc. First Int. Cong. Ecol.*, pp. 405-408.

GOMATAM, J. (1974a): A new model for interacting populations. I: Two-species systems, *Bull. Math. Biol.* 36:347-353.

______(1974b): A new model for interacting populations. II: Principle of competitive exclusion, *Bull. Math. Biol.* 36:355 365.

GOMATAM, J., and N. MacDONALD (1975): Time delays and stability of two competing species, *Math. Biosci.* 24:247-255.

GOPALSAMY, K. (1976): Random population clusters and transport, *Math. Biosci.* 29:259-272.

______(1977a): Competition, dispersion and coexistence, *Math. Biosci.* 33:25-33.

______(1977b): Competition and coexistence in spatially heterogeneous environments, *Math. Biosci.* 36:229-242.

GOPALSAMY, K., and B. D. AGGARWALA (1980): Recurrence in two-species competition, *Ecol. Model.* 9:153-163.

GOSWAMI, D., and A. LAHIRI (1979): A generalized prey-predator model, *J. Theor. Biol.* 79:243-246.

GOUDRIAAN, J., and C. T. de WIT (1973): A re-interpretation of Gause's population experiments by means of simulation, *J. Anim. Ecol.* 42:521-530.

GOULDEN, C. E., and L. L. HORNIG (1980): Population oscillations and energy reserves in planktonic eladocera and their consequences to competition, *Proc. Nat. Acad. Sci.* USA 77: 1716-1720.

GOURLEY, R. S., and C. E. LAWRENCE (1977): Stable population analysis in periodic environments, *Theor. Pop. Biol.* 11:49-59.

GRANERO PORATI, M. I., A. PORATI, and A. VECLI (1978): Analytical conditions for the conservative form of the ecological equations, *Bul. Math. Biol.* 40:257-264.

GRANT, P. R. (1978): Dispersal in relation to carrying capacity, *Proc. Nat. Acad. Sci.* USA 75:2854-2858.

GRASMAN, J., and E. VELING (1973): An asymptotic formula for the period of a Volterra-Lotka system, *Math. Biosci.* 18:185-189.

GRASMAN, W. (preprint): Finding periodic solutions in autonomous differential equations in higher dimensional spaces.

GREEN, R. H. (1974): Multivariate niche analysis with temporally varying environmental factors, *Ecology* 55:73-83.

GRENNEY, W. J., D. A. BELLA, and H. C. CURL, Jr. (1973): A theoretical approach to interspecific competition in phytoplankton communities, *Amer. Nat.* 107:405-425.

GRIFFEL, D. H. (1976): Age-dependent population growth, *J. Inst. Math. Appl.* 17:141-152.

______(1979): Harvesting competing populations, *Rocky Mountain J. Math.* 9:87-91.

GRINNELL, J. (1904): The origin and distribution of the chestnut-backed chickadee, *Auk* 21:364-382.

GROSSBERG, S. (1978a): Decisions, patterns, and oscillations in nonlinear competitive systems with applications to Volterra-Lotka systems, *J. Theor. Biol.* 73:101-130.

______(1978b): Competition, decision and consensus, *J. Math. Anal. Appl.* 66:470-493.

GRUBER, B. (1976): The influence of saturation on the predator-prey relations, *Theor. Pop. Biol.* 10:173-184.

GUCKENHEIMER, J., G. OSTER, and A. IPAKTCHI (1977): The dynamics of density dependent population models, *J. Math. Biol.* 4:101-147.

GULMAHAMAD, H., and P. DE BACH (1978): Biological control of the San Jose scale Quadraspidiotus perniciosus (Comstock) (Homoptera: Diaspididae) in southern California, *Hilgardia* 96:205-238.

GURNEY, W. S. C. and R. M. NISBET (1978a): Predator-prey fluctuations in patchy environments, *J. Anim. Ecol.* 47:85-102.

______(1978b): Single-species population fluctuations in patchy environments, *Amer. Nat.* 112:1075-1090.

______(1979): Ecological stability and social hierarchy, *Theor. Pop. Biol.* 16:48-80.

GURTIN, M. E. (1974): Some mathematical models for population dynamics that lead to segregation, *Quart. J. Appl Math.* 32: 1-9.

GURTIN, M. E., and D. S. LEVINE (1979): On predator-prey interactions with predation dependent on age of prey, *Math. Biosci.* 47:207-219.

GURTIN, M. E., and R. C. MacCAMY (1974): Non-linear age-dependent population dynamics, *Arch. Rat. Mech. Anal.* 54:281-300.

______(1979): Some simple models for nonlinear age-dependent population dynamics, *Math. Biosci.* 43:199-211.

HABTE, M. and M. ALEXANDER (1978): Protozoan density and the coexistence of protozoan predators and bacterial prey, *Ecology* 59:140-146.

HADELER, K. P. (1976): On the stability of the stationary state of a population growth equation with time-lag, *J. Math. Biol.* 3: 197-201.

HADELER, K. P., and U. LIBERMAN (1975): Selection models with fertility difference, *J. Math. Biol.* 2:19-32.

HADELER, K. P., and J. TOMIUK (1977): Periodic solutions of difference equations, *Arch. Rat. Mech. Anal.* 65:87-95.

HAIGH, J., and J. MAYNARD SMITH (1972): Can there be more predators than prey? *Theor. Pop. Biol.* 3:290-299.

HAIRSTON, N. G., F. E. SMITH, and L. B. SLOBODKIN (1960): Community structure, population control, and competition, *Amer. Nat.* 94:420-425.

HALFON, E., and M. G. REGGIANI (1978): Adequacy of ecosystem models, *Ecol. Model* 4:41-50.

HALLAM, T. G. (1978): Structural sensitivity of grazing formulations in nutrient controlled plankton models, *J. Math. Biol.* 5:269-280.

______(preprint): On persistence of aquatic ecosystems.

HALLAM, T. G., and D. S. SIMBERLOFF (1977): On the intrinsic structure of differential equation models of ecosystems, *Ecol. Model.* 3:167-182.

HALLAM, T. G., L. J. SVOBODA, and T. C. GARD (1979): Persistence and extinction in three species Lotka-Volterra competitive systems, *Math. Biosci.* 46:117-124.

HALLETT, J. G., and S. L. PIMM (1979): Direct estimation of competition, *Amer. Nat.* 113:593-600.

HAMANN, J. R., and L. M. BIANCHI (1970): Stochastic population mechanics in the relational systems formalism: Volterra-Lotka ecological dynamics, *J. Theor. Biol.* 28:175-184.

HAMILTON, W. D., and R. M. MAY (1977): Dispersal in stable habitats, *Nature* 269:578-581.

HANNON, B. (1979): Total energy costs in ecosystems, *J. Theor. Biol.* 80:271-293.

HANSKI, I. (1978): Some comments on the measurement of niche metrics, *Ecology* 59:168-174.

HARADA, K., and T. FUKAO (1978): Coexistence of competing species over a linear habitat of finite length, *Math. Biosci.* 38: 279-291.

HARDMAN, J. M. (1976a): Life table data for use in deterministic and stochastic simulation models predicting the growth of insect populations under malthusian conditions, *Can. Ent.* 108:897-906.

______(1976b): Deterministic and stochastic models simulating the growth of insect populations over a range of temperatures under malthusian conditions, *Can. Ent.* 108:907-924.

HARNER, E. J., and R. C. WHITMORE (1977): Multivariate measures of niche overlap using discriminant analysis, *Theor. Pop. Biol.* 12:21-36.

HARPENDING, H. C. (1979): The population genetics of interactions, *Amer. Nat.* 113:622-630.

HARRIS, G., and I. OLSEN (1976): A possible mechanism for the synthesis of antibodies, *J. Theor. Biol.* 58:417-423.

HARRIS, J. R. W. (1974): The kinetics of polyphagy, in M. B. Usher and M. H. Williamson, eds., *Ecological Stability*, Chapman and Hall, London, pp. 123-139.

HARRISON, G. W. (1979a): Stability under environmental stress: resistance, resilience, persistence, and variability, *Amer. Nat.* 113:659-669.

______(1979b): Global stability of predator-prey interactions, *J. Math. Biol.* 8:159-171.

______(1979c): Global stability of food chains, *Amer. Nat.* 114: 455-457.

______(to appear): Persistence of predator-prey systems in an uncertain environment, *J. Math. Biol.*

HARTMAN, P. (1973): *Ordinary Differential Equations*, P. Hartman, Baltimore.

HARWELL, M. A., W. P. CROPPER, JR., and H. L. RAGSDALE (1977): Nutrient recycling and stability: a reevaluation, *Ecology* 58:660-666.

HASSELL, M. P. (1969): A population model for the interaction between *Cyzenis albicans* (Fall.) (Tachinidae) and *Operophtera brumata* (L.) (Geometridae) at Wythan, Berkshire, *J. Anim. Ecol.* 38:567-576.

______(1971): Mutual interference between searching insect parasites, *J. Anim. Ecol.* 40:473-486.

______(1975): Density-dependence in single-species populations, *J. Anim. Ecol.* 44:283-295.

______(1978): *The Dynamics of Arthropod Predator-prey Systems*, Princeton University Press, Princeton.

HASSELL, M. P., and H. N. COMINS (1976): Discrete time models for two-species competition, *Theor. Pop. Biol.* 9:202-221.

______(1978): Sigmoid functional responses and population stability, *Theor. Pop. Biol.* 14:62-67.

HASSELL, M. P., and C. B. HUFFAKER (1969): Regulatory processes and population cyclicity in laboratory populations of *Anagasta kuhniella* (Zeller) (Lepidoptera: Phycitidae). III: The development of population models, *Res. Pop. Ecol.* 11:186-210.

HASSELL, M. P., J. H. LAWTON, and J. R. BEDDINGTON (1976): The components of arthropod predation. I: The prey death-rate, *J. Anim. Ecol.* 45:135-164.

HASSELL, M. P., J. H. LAWTON, and R. M. MAY (1976): Patterns of dynamical behavior in single-species populations, *J. Anim. Ecol.* 45:471-486.

HASSELL, M. P., and R. M. MAY (1973): Stability in insect host-parasite models, *J. Anim. Ecol.* 42:693-726.

______(1974): Aggregation of predators and insect parasites and its effect on stability, *J. Anim. Ecol.* 43:567-594.

HASSELL, M. P., and D. J. ROGERS (1972): Insect parasite responses in the development of population models, *J. Anim. Ecol.* 41: 661-676.

HASSELL, M. P., and G. C. VARLEY (1969): New inductive population model for insect parasites and its bearing on biological control, *Nature* 223:1133-1137.

HASTINGS, A. (1977): Spatial heterogeneity and the stability of predator-prey systems, *Theor. Pop. Biol.* 12:37-48.

______(1978a): Global stability in two species systems, *J. Math. Biol.* 5:399-403.

______(1978b): Global stability in Lotka-Volterra systems with diffusion, *J. Math. Biol.* 6:163-168.

______(1978c): Spatial heterogeneity and the stability of predator-prey systems: predator-mediated coexistence, *Theor. Pop. Biol.* 14:380-395.

HASTINGS, H. M. (1979): Stability considerations in community organization, *J. Theor. Biol.* 78:121-127.

HASTINGS, H. M., and M. CONRAD (1979): Length and evolutionary stability of food chains, *Nature* 282:838-839.

HAUSRATH, A. R. (1975): Stability properties of a class of differential equations modelling predator-prey relationships, *Math. Biosci.* 26:267-281.

HAUSSMANN, U. G. (1971): Abstract food webs in ecology, *Math. Biosci.* 11:291-316.

______(1973): On the principle of competitive exclusion, *Theor. Pop. Biol.* 4:31-41.

______(1974): Coexistence of species in a discrete system, in P. Van den Driessche, ed., *Mathematical Problems in Biology*, Springer-Verlag Lecture Notes in Biomathematics, Vol. #2, New York, pp. 73-82.

HAYNES, D. L., and P. SISOJEVIC (1966): Predatory behaviour of *Philodromus rufus* Walckenaer (Araneae: Themisidae), *Can. Ent.* 98:113-133.

HELLER, R. (1978): Two predator-prey difference equations considering delayed population growth and starvation, *J. Theor. Biol.* 70:401-413.

HILBORN, R. (1975): The effect of spacial heterogeneity on the persistence of predator-prey interactions, *Theor. Pop. Biol.* 8:346-355.

______(1979): Some long term dynamics of predator-prey models with diffusion, *Ecol. Model.* 6:23-30.

HINES, W. G. S. (1977): Competition with an evolutionary stable strategy, *J. Theor. Biol.* 67:141-153.

HIRATA, H. (1980): A model of hierarchial ecosystems with migration, *Bul. Math. Biol.* 42:119-130.

HIRSCH, M. W., and S. SMALE (1974): *Differential Equations, Dynamical Systems, and Linear Algebra*, Academic Press, New York.

HOFBAUER, J., P. SCHUSTER, and K. SIGMUND (1979): A note on evolutionary stable strategies and game dynamics, *J. Theor. Biol.* 81:609-612.

HÖGSTEDT, G. (1980): Prediction and test of the effects of interspecific competition, *Nature* 283:64-66.

HOLLING, C. S. (1959): Some characteristics of simple types of predation and parasitism, *Can. Ent.* 91:385-398.

______(1968): The tactics of a predator, *Symp. R. Ent. Soc. Lond.* 4:47-58.

HOLT, R. D. (1977): Predation, apparent competition, and the structure of prey communities, *Theor. Pop. Biol.* 12:197-229.

HOPPENSTEADT, F. C., and J. M. HYMAN (1977): Periodic solutions of a logistic difference equation, *SIAM J. Appl. Math.* 32: 73-81.

HORN, H. S., and R. H. MacARTHER (1972): Competition among fugitive species in a harlequin environment, *Ecology* 53:749-752.

HOUSTON, D. B. (1978): Elk as winter-spring food for carnivours in northern Yellowstone National Park, *J. Appl. Ecol.* 15: 653-661.

HSU, I.-D., and N. D. KAZARINOFF (1977): Existence and stability of periodic solutions of a third order nonlinear autonomous system simulating an immune response in animals, *Proc. R. Soc. Edinb.* A77:163-175.

HSU, S. B. (1976): A Mathematical Analysis of Competition for a Single Resource, Ph. D. Thesis, University of Iowa, Iowa City.

______(1978a): Limiting behaviour for competing species, *SIAM J. Appl. Math.* 34:760-763.

______(1978b): The application of the Poincaré transform to the Lotka-Volterra model, *J. Math. Biol.* 6:67-73.

______(1978c): On global stability of a predator-prey system, *Math. Biosci.* 39:1-10.

______(preprint, a): On a periodic population equation.

______(preprint, b): A competition model for a seasonally fluctuating nutrient.

______(preprint, c): On a resource based ecological competition model with interference.

HSU, S. B., and K. S. CHENG (preprint): Exploitative competition of two micro-organisms for two complementary nutrients in continuous culture.

HSU, S. B., and S. P. HUBBELL (1979): Two predators competing for two prey species: an analysis of MacArthur's model, *Math. Biosci.* 47:143-171.

HSU, S. B., S. HUBBELL, and P. WALTMAN (1977): A mathematical theory for single-nutrient competition in continuous cultures of micro-organisms, *SIAM J. Appl. Math.* 32:366-383.

______(1978): Competing predators, *SIAM J. Appl. Math.* 35:617-625.

HUBBELL, S. P. (1973a): Populations and simple food webs as energy filters. I: One-species systems, *Amer. Nat.* 107:94-121.

______(1973b): Populations and simple food webs as energy filters. II: Two-species systems, *Amer. Nat.* 107:122-151.

HUBBELL, S. P., and S. B. HSU (preprint): Exploitative competition for resources: multiple mechanistic origins for each outcome of classical competition theory.

HUBBELL, S. P., and P. A. WERNER (1979): On measuring the intrinsic rate of increase of populations with heterogeneous life histories, *Amer. Nat.* 113:277-293.

HUDDLESTON, J. V., C. G. De WALD, and H. N. JAGADAESH (1974): A dynamic model of an environmental system with n interacting components and p degrees of freedom, *Bull. Math. Biol.* 36: 91-96.

HUFFAKER, C. B., and C. E. KENNETT (1956): Experimental studies on predation: predation and cyclamen-mite populations on strawberries in California, *Hilgardia* 26:191-222.

HUFFAKER, C. B., and J. E. LAING (1972): "Competitive Displacement" without a shortage of resources? *Res. Pop. Ecol.* 14:1-17.

HURD, L. E., and L. L. WOLF (1974): Stability in relation to nutrient enrichment in arthopod consumers of old-field successional ecosystems, *Ecol. Monog.* 44:465-482.

HURLBERT, S. H. (1978): The measurement of niche overlap and some relatives, *Ecology* 59:67-77.

HUSTON, M. (1979): A general hypotheses of species diversity, *Amer. Nat.* 113:81-101.

HUTCHINSON, G. E., (1957): Concluding remarks, *Cold Spring Harbor Symp. on Quant. Biol.* 25:415-427.

______(1961): The paradox of the plankton, *Amer. Nat.* 95:137-145.

IKEDA, M., and D. D. SILJAK (1980): Lotka-Volterra equations: decomposition, stability, and structure part I: equilibrium analysis, *J. Math. Biol.* 9:65-83.

INNIS, G. (1972): The second derivative and population modelling: another view, *Ecology* 53:720-723.

INOUYE, R. S. (1980): Stabilization of a predator-prey equilibrium by addition of a second "keystone" victim, *Amer. Nat.* 115: 300-305.

ISHIHARA, M., K. HOZUMI, and K. SHINOZAKI (1972): A mathematical model of the food-consumer system. I: A case without food replenishment, *Res. Pop. Ecol.* 13:114-126.

ISTOCK, C. A. (1977): Logistic interaction of natural populations of two species of waterboatmen, *Amer. Nat.* 111:279-287.

ITO, Y. (1971): Some notes on the competitive exclusion principle, *Res. Pop. Ecol.* 13:46-54.

IWASA, Y. (1978): Optimal death strategy of animal populations, *J. Theor. Biol.* 72:611-626.

JAEGER, R. G. (1972): Food as a limited resource in competition between two species of terrestial salamanders, *Ecology* 53: 535-546.

______(1974): Competitive exclusion: comments on survival and extinction of species, *Bioscience* 24:33-39.

JAROSZEWSKI, J., A. AHMED, and K. W. SELL (1976): Mechanism for regulation of immune responses, *J. Theor. Biol.* 57:121-129.

JEFFRIES, C. (1974): Qualitative stability and digraphs in model ecosystems, *Ecology* 55:1415-1419.

______(1975): Stability of ecosystems with complex food webs, *Theor. Pop. Biol.* 7:145-155.

______Probabilistic limit cycles, in P. Van den Driessche, ed., *Mathematical Problems in Biology*, Springer-Verlag Lecture Notes in Biomathematics, Vol. 2, New York, pp. 123-131.

______(1976): Stability of predation ecosystem models, *Ecology* 57:1321-1325.

JENSON, A. L., and R. C. BALL (1970): Variation in the availability of food as a cause of fluctuations in predator and prey population densities, *Ecology* 51:517-520.

JERNE, N. K. (1973): The immune system, *Sci. Amer.* 229:52-60.

______(1974): The immune system: a web of V domains, *The Harvey Lect.* 70:93-110.

JOHNSON, R. H. (1910): Determinate evolution in color pattern of the lady-beetles, *Carnegie Inst. Wash, Publ.* 122:104 pp.

JONES, D. D., and C. T. Walters (1976): Catastrophy theory and fisheries regulation, *J. Fish. Res. Board Can.* 33:2829-2833.

JONES, J. M. (1976): The r-K selection continuum, *Amer. Nat.* 110: 320-323.

JORNE, J. (1977): The diffusive Lotka-Volterra oscillating system, *J. Theor. Biol.* 65:133-139.

JORNE, J., and S. Carmi (1977): Liapunov stability of the diffusive Lotka-Volterra Equations, *Math. Biosci.* 37:51-61.

KANNAN, D. (1976): On some Markov models of certain interacting populations, *Bull. Math. Biol.* 38:723-738.

______(in press): Volterra Verhulst prey predator systems with time dependent coefficients: diffusion type approximation and periodic solutions, *Bull. Math. Biol.*

KAPLAN, J. L., and J. A. YORKE (1975): On the stability of a periodic solution of a differential delay equation, *SIAM J. Appl. Math.* 6:268-282.

______(1977): Competitive exclusion and nonequilibrium coexistence, *Amer. Nat.* 111:1030-1036.

KAWASAKI, K., and E. TERAMOTO (1979): Spatial pattern formation of prey-predator populations, *J. Math. Biol.* 8:33-46.

KAZARINOFF, N. D., and P. VAN DEN DRIESSCHE (1978): A model predator-prey system with functional response, *Math. Biosci.* 39:125-134.

______(1979): Control of oscillations in hematopoiesis, *Science* 203:1348-1349.

KAZARINOFF, N. D., Y.-H. WAN and P. VAN DEN DRIESSCHE (1978): Hopf bifurcation and stability of periodic solutions of differential-difference and integro-differential equations, *J. Indust. Maths. Applics.* 21:461-477.

KEENER, J. P. (1979): Long range predation in pattern formation, *Rocky Mountain J. Math.* 9:99-113.

KEIDING, N. (1975): Extinction and exponential growth in random environments, *Theor. Pop. Biol.* 8:49-63.

KEMPTON, R. A. (1979): The structure of species abundance and measurement of diversity, *Biometrics* 35:307-321.

KERNER, E. H. (1961): On the Volterra-Lotka principle, *Bull. Math. Biophys.* 23:141-157.

______(1971): Statistical-mechanical theories in biology, *Adv. in Chem. Phys.* 19:325-352.

KHANIN, M. A., and N. L. DORFMAN (1973): Mathematical model in growth based on the evolutionary principle of extremes, *Dokl. Bio. Sci.* 212:371-374.

KIESTER, A. R., and R. BARAKAT (1974): Exact solutions to certain stochastic differential models of population growth, *Theor. Pop. Biol.* 6:199-216.

KILMER, W. L. (1972): On some realistic constraints in prey-predator mathematics, *J. Theor. Biol.* 36:9-22.

KILMER, W. L., and T. H. PROBERT (1977): Oscillatory depletion models for renewable ecosystems, *Math. Biosci.* 36:25-29.

KIRITANI, K., and N. KAKIYA (1975): An analysis of the predator-prey system in the paddy field, *Res. Pop. Ecol.* 17:29-38.

KNOLLE, H. (1976): Lotka-Volterra equations with time delay and periodic forcing term, *Math. Biosci.* 31:351-375.

KOCH, A. L. (1974a): Coexistence resulting from an alteration of density dependent and density independent growth, *J. Theor. Biol.* 44:373-386.

______(1974b): Competitive coexistence of two predators utilizing the same prey under constant environmental conditions, *J. Theor. Biol.* 44:387-395.

KOLATA, G. B. (1975): Cascading bifurcations: the mathematics of chaos, *Science* 189:984-985.

KOLMOGOROV, A. (1936): Sulla teoria di Volterra della lotta per l'esistenza, *Gi. Inst. Ital. Attuari* 7:74-80.

KRAPIVIN, V. F. (1972): Investigation of generalized predator-prey model, *Sov. J. Ecol.* 3:215-221.

KRIKORIAN, N. (1979): The Volterra model for three species predator-prey systems: boundedness and stability, *J. Math. Biol.* 7:117-132.

KRISHNAMURTHY, L. (1977): Functional properties and organization of grazing lands ecosystem, *J. Theor. Biol.* 68:65-72.

KROES, H. W. (1977): The niche structure of ecosystems, *J. Theor. Biol.* 65:317-326.

LADDE, G. S. (1976): Stability of model ecosystems with time-delay, *J. Theor. Biol.* 61:1-13.

LADDE, G. S., and D. D. SILJAK (1975): Stability of multispecies communities in randomly varying environments, *J. Math. Biol.* 2:165-178.

LAKIN, W. D., and P. VAN DEN DRIESSCHE (1977): Time scales in population biology, *SIAM J. Appl. Math.* 32:694-705.

LANDAHL, H. D., and B. D. HANSEN (1975): A three stage population model with cannibalism, *Bull. Math. Biol.* 37:11-17.

LAPLANTE, J.-P. (1979): Inhomogeneous steady state distributions of species in predator-prey systems. A specific example, *J. Theor. Biol.* 81:29-45.

LA SALLE, J., and S. LEFSCHETZ (1961): *Stability by Liapunov's Direct Method*, Academic Press, New York.

LAW, R. (1979): Optimal life histories under age-specific predation, *Amer. Nat.* 114:399-417.

LAWLOR, L. R., and J. MAYNARD SMITH (1976): The coevaluation and stability of competing species, *Amer. Nat.* 110:79-99.

LAWTON, J. H., J. R. BEDDINGTON, and R. BONSER (1974): Switching in invertebrate predators, in M. B. Usher and M. H. Williamson, eds., *Ecological Stability*, Chapman and Hall, London, pp. 141-158.

LAWTON, J. H., M. P. HASSELL, and J. R. BEDDINGTON (1975): Prey death rates and rate of increase of anthropod predator populations, *Nature* 255:60-62.

LEE, K. Y., R. O. BARR, S. H. GAGE, and A. N. KHARKAR (1976): Formulation of a mathematical model for insect pest ecosystems-the cereal leaf beetle problem, *J. Theor. Biol.* 59:33-76.

LEFEVER, R., and W. HORSTHEMKE (1979): Bistability in fluctuating environments. Implications in tumor immunology, *Bull. Math. Biol.* 41:469-490.

LEFKOVITCH, L. P. (1966): A population growth model incorporating delayed responses, *Bul. Math. Biophys.* 28:219-233.

LEIGH, E. R. (1968): The ecological role of Volterra's equations, in M. Gerstenhaber, ed., *Some Mathematical Problems in Biology*, American Mathematical Society, Providence, pp. 1-61.

LEON, J. A. (1975): Limit cycles in populations with separate generations, *J. Theor. Biol.* 49:241-244.

LEON, J. A., and D. B. TUMPSON (1975): Competition between two species for two complementary or substitutable resources, *J. Theor. Biol.* 50:185-201.

LERNER, I. M., and E. R. DEMPTER (1962): Indeterminism in interspecific competition, *Proc. Nat. Acad. Sci. USA* 48:821-826.

LESLIE, P. H. (1957): An analysis of the data for some experiments carried out by Gause with populations of the protozoa, *Paramecium aurelia* and *Paramecium caudatum*, *Biometrika* 44: 314-327.

______(1958): A stochastic model for studying the properties of certain biological systems by numerical methods, *Biometrika* 45:16-31.

LESLIE, P. H., and J. C. GOWER (1958): The properties of a stochastic model for two competing species, *Biometrika* 45:316-330.

______(1960): The properties of a stochastic model for the predator-prey type of interaction between two species, *Biometrika* 47:219-234.

LESLIE, P. H., T. PARK, and D. B. MERTZ (1968): The effect of varying the initial numbers on the outcome of competition between two Tribolium species, *J. Anim. Ecol.* 37:9-23.

LEUNG, A. (1976): Limiting behavior for several interacting populations, *Math. Biosci.* 29:85-98.

______(1977): Periodic solutions for a prey-predator differential delay equation, *J. Differ. Equations* 26:391-403.

______(1979): Conditions for global stability concerning a prey-predator model with delay effects, *SIAM J. App. Math.* 36: 281-286.

______(1980): Equilibria and stabilities for competing-species reaction-diffusion equations with Dirichlet boundary data, *J. Math. Anal. Appl.* 73:204-218.

LEVANDOWSKY, M. (1972): Ecological niches of sympatric phytoplankton species, *Amer. Nat.* 106:71-78.

LEVIN, B. R., F. M. STEWART, and L. CHAO (1977): Resource-limited growth, competition, and predation: a model and experimental studies with bacteria and bacteriophage, *Amer. Nat.* 111:3-24.

LEVIN, S. A. (1970): Community equilibria and stability, and an extension of the competitive exclusion principle, *Amer. Nat.* 104:413-423.

______(1972): A mathematical analysis of the genetic feedback mechanism, *Amer. Nat.* 106:145-164.

______(1974): Dispersion and population interactions, *Amer. Nat.* 108:207-228.

______(1975a): ed. *Ecosystem Analysis and Prediction*, SIAM, Philadelphia.

______(1975b): On the care and use of mathematical models, *Amer. Nat.* 109:785-786.

______(1976): Spatial patterning and the structure of ecological communities, in *Some Mathematical Questions in Biology*, Vol. VII, American Mathematical Society, Providence.

______(1977): A more functional response to predator-prey stability, *Amer. Nat.* 111:381-383.

______(1978): Population models and community structure, in S. A. Levin, ed., *Studies in Mathematical Biology Part II: Populations and Communities*, Mathematical Association of America, Washington: 439-476.

______(preprint): Population dynamic models in heterogeneous environments.

LEVIN, S. A., and R. M. MAY (1976): A note on difference-delay equations, *Theor. Pop. Biol.* 9:178-187.

LEVIN, S. A., and L. A. SEGEL (1976): Hypothesis for the origin of plankton patchiness, *Nature* 259:659.

LEVIN, S. A., and J. D. UDOVIC (1977): A mathematical model of coevolving populations, *Amer. Nat.* 111:657-675.

LEVINE, D. S. (1979): Existence of limiting pattern for a system of nonlinear equations describing interpopulation competition, *Bul. Math. Biol.* 41:617-628.

LEVINE, S. H. (1975): Discrete time modeling of ecosystems with applications in environmental enrichment, *Math. Biosci.* 24: 307-317.

LEVINE, S. H., F. M. SCUDO, and D. J. PLUNKETT (1977): Persistence and convergence of ecosystems: an analysis of some second order difference equations, *J. Math. Biol.* 4:171-182.

LEVINS, R. (1968): *Evolution in Changing Environments*, Princeton University Press, Princeton, N. J.

______(1969): The effect of random variations of different types on population growth, *Proc. Nat. Acad. Sci. USA* 62:1061-1065.

LEVINS, R., and D. CULVER (1971): Regional coexistence of species and competition between rare species, *Proc. Nat. Acad. Sci. USA* 68:1246-1248.

LEVINTON, J. S. (1979): A theory of diversity equilibrium and morphological evolution, *Science* 204:335-336.

LEWIS, E. R. (1972): Delay-line models of population growth, *Ecology* 53:797-807.

______(1977): Linear population models with stochastic time delay, *Ecology* 58:738-749.

LEWONTIN, R. C. (1969): The meaning of stability, in *Diversity and Stability in Ecological Systems*, Brookhaven Symposia in Biology, Vol. 22, pp. 13-24.

LEWONTIN, R. C., and D. COHEN (1969): On population growth in a randomly varying environment, *Proc. Nat. Acad. Sci. USA* 62:1056-1060.

LI, T-Y., and J. A. YORKE (1975): Period three implies chaos, *Amer. Math. Mon.* 82:985-992.

LIDICKER, W. Z., Jr. (1962): Emigration as a possible mechanism permitting the regulation of population density below carrying capacity, *Amer. Nat.* 96:29-33.

LIN, J., and P. B. KAHN (1976): Averaging methods in predator-prey systems and related biological models, *J. Theor. Biol.* 57:73-102.

______(1977a): Qualitative behavior of predator-prey communities, *J. Theor. Biol.* 65:101-132.

______(1977b): Limit cycles in random environments, *SIAM J. Appl. Math.* 32:260-291.

______(1978): Qualitative dynamics of three species predator-prey systems, *J. Math. Biol.* 5:257-268.

______(preprint): Qualitative behavior of predator-prey communities.

LLOYD, M., and J. WHITE (1980): On reconciling patchy microspatial distributions with competition models, *Amer. Nat.* 115:29-44.

LOGAFET, D. O. (1975a): On the stability of a class of matrices arising in the mathematical theory of biological associations, *Sov. Math. Dokl.* 16:523-527.

______(1975b): Investigation of systems of n pairs of "predator-prey," connected through competition, *Dokl. Akad. Nauk SSSR* 224:529-531.

LOMNICKI, A. (1978): Individual differences between animals and the natural regulation of their numbers, *J. Anim. Ecol.* 47: 461-475.

LONG, G. E., P. H. DURAN, R. O JEFFORDS, and D. N. WELDON (1974): An application of the logistic equation to the population dynamics of salt-marsh gastropods, *Theor. Pop. Biol.* 5:450-459.

LOTKA, A. J. (1925): *Elements of Physical Biology*, Williams and Wilkins, Baltimore.

LOUD, W. S. (1959): Periodic solutions of $x'' + cx' + g(x) = \varepsilon f(t)$, *Mem. Amer. Math. Soc.* No. 31.

______Periodic solutions of perturbed second-order autonomous equations, *Mem. Amer. Math. Soc.* No. 47.

LUCK, R. F., J. C. VAN LENTEREN, P. H. TWINE, L. KUENEN, and T. UNRUH (1979): Prey or host searching behavior that leads to a sigmoid functional response in invertebrate predators and parasitoids, *Res. Pop. Ecol.* 20:257-264.

LUCKINBILL, L. S. (1973): Coexistence in laboratory populations of Paramecium aurelia and its predator Didinium nasutum, *Ecology* 54:1320-1327.

______(1974): The effects of space and enrichment on a predator-prey system, *Ecology* 55:1142-1147.

______(1979): Selection and the r/K continuum in experimental populations of protozoa, *Amer. Nat.* 113:427-437.

LUDWIG, D. (1975): Persistence of dynamical systems under random perturbations, *SIAM Rev.* 17:605-640.

______(1978): Comparison of some deterministic and stochastic population theories, in S. A. Levin, ed., *Studies in Mathematical Biology Part II: Populations and Communities,* Mathematical Association of America, Washington:367-388.

LUDWIG, D., D. G. ARONSON, and H. F. WEINBERGER (1979): Spatial patterning of the spruce budworm, *J. Math. Biol.* 8:217-258.

LUDWIG, D., D. D. JONES, and C. S. HOLLING (1978): Qualitative analysis of insect outbreak systems: the spruce budworm and forest, *J. Anim. Ecol.* 47:315-332.

MACARTHUR, R. (1968a): Selection for life tables in periodic environments, *Amer. Nat.* 102:381-383.

______(1968b): The theory of the niche, in R. C. Lewontin, ed., *Population Biology and Evolution*, Syracuse University Press.

______(1969): Species packing, and what interspecies competition minimizes, *Proc. Nat. Acad. Sci. USA* 64:1369-1371.

MACARTHUR, R., and R. LEVINS (1967): The limiting stability, convergence, and divergence of coexisting species, *Amer. Nat.* 101:377-385.

MACDONALD, N. (1975): The stability of a feasible random ecosystem, *Math. Biosci.* 27:141-143.

______(1976a): Time delay in predator-prey models, *Math. Biosci.* 28:321-330.

______(1976b): Extended diapause in a discrete generation population model, *Math. Biosci.* 31:255-257.

______(1978): *Time Lags in Biological Models*, Lecture Notes in Biomathematics, No. 27, Springer-Verlag, New York.

______(1979): Simple aspects of foodweb complexity, *J. Theor. Biol.* 80:577-588.

MAGUIRE, B., Jr. (1967): A partial analysis of the niche, *Amer. Nat.* 101:515-523.

______(1973): Niche response structure and the analytical potentials of its relationship to the habitat, *Amer. Nat.* 107: 213-246.

MALY, E. J. (1975): Interactions among the predatory rotifer Asplanchna and two prey, *Paramecium* and *Euglena*, *Ecology* 56: 346-358.

______(1978): Stability of the interaction between Didinium and Paramecium: effects of dispersal and predator time lag, *Ecology* 59:733-741.

MANGEL, M., and D. LUDWIG (1977): Probability of extinction in a stochastic environment, *SIAM J. Appl. Math.* 33:256-266.

MANTON, M. J. (1979): On feasible random ecosystems, *Bul. Math. Biol.* 41:751-755.

MARGALEF, R. (1969): Diversity and stability: a practical proposal and a model of interdependence, in *Diversity and Stability in Ecological Systems*, Brookhaven Symposia in Biology, Vol. 22, pp. 25-37.

MARSDEN, J. E., and M. McMCRACKEN (1976): *The Hopf Bifurcation and Its Applications,* Springer-Verlag, New York.

MARSOLAN, N. F., and W. G. RUDD (1976): Modeling and optimal control of insect pest populations, *Math. Biosci.* 30:231-244.

MARTEN, G. G. (1973): An optimization equation for predation, *Ecology* 54:92-101.

MATSUMOTO, B. M., and C. B. HUFFAKER (1973a): Regulatory processes and population cyclicity in laboratory populations of *Anagasta kuhniella* (Zeller) (Lepidoptera: Physitidae). IV: The sequential steps in the behavioral process of host finding and parasitization by the entomophagous parasite, *Venturia canescens* (Gravenhorst) (Hymenoptera: Ichneumonidae) *Res. Pop. Ecol.* 15:23-31.

______(1973b): Regulatory processes and population cyclicity in laboratory populations of *Anagasta kuhniella* (Zeller) (Lepidoptera: Physitidae). V: Host finding and parasitization in a "small" universe by an entomophagous parasite, *Venturia canescens* (Gravenhorst) (Hymenoptera: Ichneumonidae) *Res. Pop. Ecol.* 15:32-49.

______(1974): Regulatory processes and population cyclicity in laboratory populations of *Anagasta kuhniella* (Zeller) (Lepidoptera: Physitidae). VI: Host finding and parasitization in a "large" universe by an entomophagous parasite, *Venturia canescens* (Gravenhorst) (Hymenoptera: Ichneumonidae) *Res. Pop. Ecol.* 15:193-212.

MAY, R. M. (1972): Limit cycles in predator-prey communities, *Science* 177:900-902.

______(1973a): *Stability and Complexity in Model Ecosystems*, Princeton University Press.

______(1973b): On relationships among various types of population models, *Amer. Nat.* 107:46-57.

______(1973c): Mass and energy flow in closed ecosystems: a comment, *J. Theor. Biol.* 39:155-163.

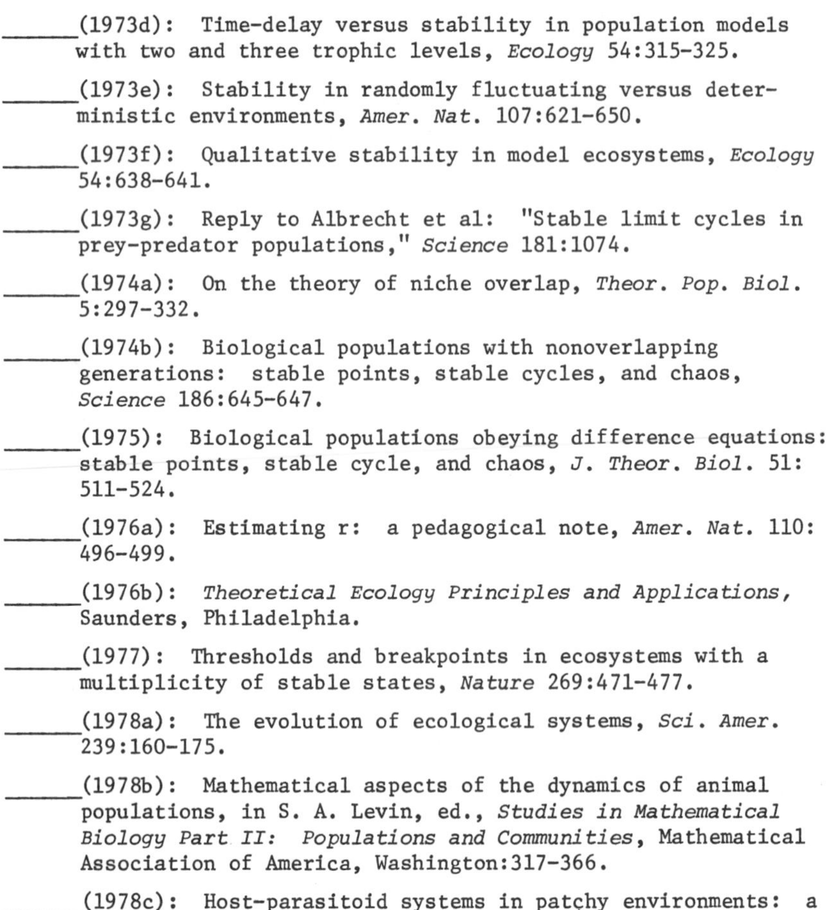

______(1973d): Time-delay versus stability in population models with two and three trophic levels, *Ecology* 54:315-325.

______(1973e): Stability in randomly fluctuating versus deterministic environments, *Amer. Nat.* 107:621-650.

______(1973f): Qualitative stability in model ecosystems, *Ecology* 54:638-641.

______(1973g): Reply to Albrecht et al: "Stable limit cycles in prey-predator populations," *Science* 181:1074.

______(1974a): On the theory of niche overlap, *Theor. Pop. Biol.* 5:297-332.

______(1974b): Biological populations with nonoverlapping generations: stable points, stable cycles, and chaos, *Science* 186:645-647.

______(1975): Biological populations obeying difference equations: stable points, stable cycle, and chaos, *J. Theor. Biol.* 51: 511-524.

______(1976a): Estimating r: a pedagogical note, *Amer. Nat.* 110: 496-499.

______(1976b): *Theoretical Ecology Principles and Applications*, Saunders, Philadelphia.

______(1977): Thresholds and breakpoints in ecosystems with a multiplicity of stable states, *Nature* 269:471-477.

______(1978a): The evolution of ecological systems, *Sci. Amer.* 239:160-175.

______(1978b): Mathematical aspects of the dynamics of animal populations, in S. A. Levin, ed., *Studies in Mathematical Biology Part II: Populations and Communities*, Mathematical Association of America, Washington:317-366.

______(1978c): Host-parasitoid systems in patchy environments: a phenomenological model, *J. Anim. Ecol.* 47:833-843.

MAY, R. M., and R. M. ANDERSON (1978): Regulation and stability of host-parasite population interactions II. Destabilizing processes, *J. Anim. Ecol.* 47:249-267.

MAY, R. M., J. R. BEDDINGTON, J. W. HORWOOD, and J. G. SHEPHERD (1978): Exploiting natural populations in an uncertain world, *Math. Biosci.* 42:219-252.

MAY, R. M., G. R. CONWAY, M. P. HASSELL, and T. R. E. SOUTHWOOD (1974): Time delays, density dependence and single-species oscillations, *J. Anim. Ecol.* 43:747-770.

MAY, R. M., and W. J. LEONARD (1975): Nonlinear aspects of competition between three species, *SIAM J. Appl. Math.* 29: 243-253.

MAY, R. M., and R. H. MACARTHUR (1972): Niche overlap as a function of environmental variability, *Proc. Nat. Acad. Sci. USA* 69:1109-1113.

MAY, R. M., and G. F. OSTER (1976): Bifurcations and dynamic complexity in simple ecological models, *Amer. Nat.* 110:573-599.

MAYNARD SMITH, J. (1968): *Mathematical Ideas in Biology*, Cambridge University Press, New York.

______(1974): *Models in Ecology*, Cambridge University Press, New York.

______(1976): What determines the rate of evolution? *Amer. Nat.* 110:331-338.

______(1977): Mathematical models in population biology, in D. E. Mathews, ed., *Mathematics and the Life Sciences*, Springer-Verlag Lecture Notes in Biomathematics, Vol. 18, New York, pp. 200-221.

MAYNARD SMITH, J., and M. SLATKIN (1973): The stability of predator-prey systems, *Ecology* 54:384-391.

MAZANOV, A. (1973a): A multi-stage population model, *J. Theor. Biol.* 39:581-587.

______(1973b): On the differential-difference growth equation, *Search* 4:199-201.

McALLISTER, C. D., R. J. LE BRASSEUR, and T. R. PARSONS (1972): Stability of enriched aquatic ecosystems, *Science* 175:562-564.

McARDLE, B. H., and J. H. LAWTON (1979): Effects of prey-size and predator-instar on the predation of Daphnia by Notonecta, *Ecol. Ent.* 4:267-275.

McGEHEE, R., and R. A. ARMSTRONG (1977): Some mathematical problems concerning the ecological principle of competitive exclusion, *J. Differ. Equations* 23:30-52.

M'CLOSKEY, R. T. (1976): Community structure in sympatric rodents, *Ecology* 57:728-739.

McMURTRIE, R. E. (1975): Determinants of stability of large randomly connected systems, *J. Theor. Biol.* 50:1-11.

______(1976): On the limit to niche overlap for nonuniform niches, *Theor. Pop. Biol.* 10:96-107.

______(1978): Persistence and stability of single-species and prey-predator systems in spatially heterogeneous environments, *Math. Biosci.* 39:11-51.

McMURTRY, J. A., and M. VAN DE VRIE (1973): Predation by *Amblyseius potentillae* (Garman) on *Panonychus ulmi* (Koch) in simple ecosystems (Acarina: Phytoseudae, Tetranychidae), *Hilgardia* 42:17-34.

McNAUGHTON, S. J. (1977): Diversity and stability of ecological communities: a comment on the role of empiricism in ecology, *Amer. Nat.* 111:515-525.

______(1978): Stability and diversity of ecological communities, *Nature* 274:251-253.

McNAUGHTON, S. J., and L. L WOLF (1970): Dominance and the niche in ecological systems, *Science* 167:131-139.

MECH, L. D. (1977): Wolf-pack buffer zones as prey reservoirs, *Science* 198:320-321.

MENDELSSOHN, R. (1978): Optimal harvesting strategies for stochastic single-species multiage class models, *Math. Biosci.* 41:159-174.

MENGE, B. A., and J. P. SUTHERLAND (1976): Species diversity gradients: synthesis of the roles of predation, competition, and temporal heterogeneity, *Amer. Nat.* 110:351-369.

MERRILL, S. J. (1976a): Mathematical models of humoral immune response, *Tech. Report*, Mathematics Department, University of Iowa, Iowa City.

______(1976b): A mathematical model of B-cell stimulation and humoral immune response, Ph. D. Thesis, University of Iowa.

______(1978a): A model of the stimulation of B-cells by replicating antigen - I, *Math. Biosci.* 41:125-141.

______(1978b): A model of the stimulation of B-cells by replicating antigen - II, *Math. Biosci.* 41:143-155.

MERTZ, D. B., D. A. CAWTHON, and T. PARK (1976): An experimental analysis of competitive indeterminacy in *Tribolium, Proc. Nat. Acad. Sci. USA* 73:1368-1372.

MERTZ, D. B., and M. J. WADE (1976): The prudent prey and the prudent predator, *Amer. Nat.* 110:489-496.

MICHALAKIS, M., C. GRUBER, and C. B. URBANI (1973): Nonlinear model of evolving populations, *Math. Biosci.* 18:269-283.

MILLER, R. S. (1964): Interspecific competition in laboratory populations of *Drosophila melanogaster* and *Drosophila simulans*, *Amer. Nat.* 98:221-238.

______(1968): Conditions of competition between redwings and yellowheaded blackbirds, *J. Anim. Ecol.* 37:43-62.

______(1969): Competition and species diversity, in *Diversity and Stability in Ecological Systems*, Brookhaven Symposia in Biology, Vol. 22, pp. 63-70.

MILLER, R. S., and D. B. BOTKIN (1974): Endangered species: models and predictions, *Amer. Sci.* 62:172-181.

MILSTEAD, W. W. (1972): Toward a quantification of the ecological niche, *Amer. Midl. Nat.* 87:346-354.

MIMURA, M. (1979): Asymptotic behaviors of a parabolic system related to planktonic prey and a predator model, *SIAM J. Appl. Math.* 37:499-512.

MIMURA, M., and T. NISHIDA (1978): On a certain semilinear parabolic system related to the Lotka-Volterra ecological model, *Publ. RIMS Kyoto Univ.* 14:269-282.

MIRMIRANI, M., and G. OSTER (1978): Competition, kin selection and evolutionary stable strategies, *Theor. Pop. Biol.* 13:304-339.

MONTROLL, E. W. (1972): Some statistical aspects of the theory of interacting species, in *Some Mathematical Questions in Biology*, Vol. III, Lectures on Mathematics in the Life Sciences, American Mathematical Society, pp. 99-143.

MOORE, R. J. (1978): Is Acanthaster planci an r-strategist? *Nature* 271:56-57.

MORRIS, H. C. (1977): Almost critical ecosystems, *Bull. Math. Biol.* 39:109-116.

MORTON, R. (1976): On the control of stochastic prey-predator models, *Math. Biosci.* 31:341-349.

MÜNSTER-SWENDSEN, M. and G. NACHANAN (1978): Asynchrony in insect host-parasite interaction and its effects on stability, studied by a simulation model, *J. Anim. Ecol.* 47:159-171.

MURDIE, G., and M. P. HASSELL (1973): Food distribution, searching success and predator-prey models, in M. S. Bartlett and R. W. Hiorns, eds., *The Mathematical Theory of the Dynamics of Biological Populations*, Academic Press, pp. 87-101.

MURDOCH, W. W. (1969): Switching in general predators: experiments on predator specificity and stability of prey populations, *Ecol. Monogr.* 39:335-354.

______(1970): Population regulation and population inertia, *Ecology* 51:497-502.

______(1977): Stabilizing effects of spatial heterogeneity in predator-prey systems, *Theor. Pop. Biol.* 11:252-273.

______(1979): Predation and the dynamics of prey populations, *Fortschr. Zool.* 25:295-310.

MURDOCH, W. W., S. AVERY, and M. E. B. SMYTH (1975): Switching in predatory fish, *Ecology* 56:1094-1105.

MURDOCH, W. W., and J. R. MARKS (1973): Predation by coccinellid beetles: experiments on switching, *Ecology* 54:160-167.

MURRAY, J. D. (1976): Spatial structures in predator-prey communities, a nonlinear time delay diffusion model, *Math. Biosci.* 30:73-85.

NEILL, W. E. (1974): The community matrix and interdependence of the competition coefficients, *Amer. Nat.* 108:399-408.

NEMYTSKII, V. V., and V. V. STEPANOV (1960): *Qualitative Theory of Differential Equations*, Princeton University Press, Princeton, N. J.

NEYMAN, J., and E. L. SCOTT (1959): Stochastic models of population dynamics, *Science* 130:303-308.

NICHOLSON, A. J. (1933): The balance of animal populations, *J. Anim. Ecol.* 2:132-178.

______(1950): Population oscillations caused by competition for food, *Nature* 165:476-477.

NICHOLSON, A. J., and V. A. BAILEY (1935): The balance of animal populations, *Proc. Zool. Soc. Lond.,* pp. 551-598.

NISBET, R. M., and W. S. C. GURNEY (1976a): Population dynamics in a periodically varying environment, *J. Theor. Biol.* 56: 459-475.

______(1976b): A simple mechanism for population cycles, *Nature* 263:319-320.

NISBET, R. M., W. S. GURNEY, and M. A. PETTIPHER (1978): Environmental fluctuations and the theory of the ecological niche, *J. Theor. Biol.* 75:223-237.

NITECKI, Z. (1978): A periodic attractor determined by one function: a counterexample to the generalized competitive exclusion principle, *J. Differential Eqns.* 29:214-234.

______(preprint): Dynamic behavior determined by one function.

NORTON, L., R. SIMON, H. D. BRERETON, and A. E. BOGDEN (1976): Predicting the course of Gompertzian growth, *Nature* 264:542-545.

NOVOTNY, J. (1976): A note on the stability of Jerne's immune network, *J. Theor. Biol.* 60:487-491.

NOY-MEIR, I. (1975): Stability of grazing systems: an application of predator-prey graphs, *J. Ecol.* 63:459-481.

______(1976): Rotational grazing in a continuously growing pasture: a simple model, *Agric. Syst.* 1:87-112.

______(1978a): Stability in simple grazing models: effects of explicit functions, *J. Theor. Biol.* 71:347-380.

______(1978b): Grazing and production in seasonal pastures: analysis of a simple model, *J. Appl. Ecol.* 15:809-835.

OATEN, A. (1977): Transit time and density dependent predation on a patchily distributed prey, *Amer. Nat.* 111:1061-1075.

OATEN, A., and W. W. MURDOCH (1975a): Functional response and stability in predator-prey systems, *Amer. Nat.* 109:289-298.

______(1975b): Switching, functional response, and stability in predator-prey systems, *Amer. Nat.* 109:299-318.

______(1977): More on functional response and stability (reply to Levin), *Amer. Nat.* 111:383-386.

ODUM, E. P. (1969): The strategy of ecosystem development, *Science* 164:262-270.

O'NEILL, R. V. (1976): Ecosystem persistence and heterotrophic regulation, *Ecology* 57:1244-1253.

OSTER, G. (1974): The role of age structure in the dynamics of interacting populations, in *Mathematical Problems in Biology*, Springer-Verlag Lecture Notes in Biomathematics, Vol. 2, pp. 166-173.

______(1976): Modelling social insect populations. I: Ergonomics of foraging and population growth in bumblebees, *Amer. Nat.* 110:215-245.

______(1978): The dynamics of nonlinear models with age structure, in S. A. Levin, ed., *Studies in Mathematical Biology Part II: Populations and Communities*, Mathematical Association of America, Washington: 411-438.

OSTER, G., A. IPAKTCHI, and S. ROCKLIN (1976): Phenotypic structure and bifurcation behavior of population models, *Theor. Pop. Biol.* 10:365-382.

OSTER, G., and Y. TAKAHASHI (1974): Models for age-specific interactions in a periodic environment, *Ecol. Monogr.* 44:483-501.

OSTER, G. F., and E. O. WILSON (1978): *Caste and Ecology in the Social Insects*, Princeton University Press.

PAINE, R. T. (1969): A note on trophic complexity and community stability, *Amer. Nat.* 103:91-93.

PAKES, A. G., A. C. TRAJSTMAN, and P. J. BROCKWELL (1979): A stochastic model for a replicating population subjected to mass emigration due to population pressure, *Math. Biosci.* 45: 137-157.

PANDE, L. K. (1978): Ecosystems with three species: one-prey-and-two-predator system in an exactly solvable model, *J. Theor. Biol.* 74:591-598.

PARK, T., and M. LLOYD (1955): Natural selection and the outcome of competition, *Amer. Nat.* 89:235-240.

PARRISH, J. D., and S. B. SAILA (1970): Interspecific competition, predation and species diversity, *J. Theor. Biol.* 27:207-220.

PATTEN, B. C. (1961): Competitive exclusion, *Science* 134:1599-1601.

______(1974): Ecosystem linearization: an evolutionary design problem, *Amer. Nat.* 109:529-539.

PAVLIDIS, T. (1973): *Biological Oscillators: Their Mathematical Analysis*, Academic Press, New York.

PEARL, R. (1930): *The Biology of Population Growth*, Knopf, New York.

PEARCE, C. (1970): A new deterministic model for the interaction between predator and prey, *Biometrics* 26:387-392.

PEARSON, E. S. (1927): The application of the theory of differential equations to the solution of problems connected with the interdependence of species, *Biometrika* 19:216-222.

PEET, R. K. (1978): Ecosystem convergence, *Amer. Nat.* 112:441-444.

PERELSON, A. S. (to appear): Optimal strategies for an immune response, in S. A. Levin, ed., *Some Mathematical Questions in Biology*, American Mathematical Society, Providence.

PERELSON, A. S., and G. I. BELL (1977): Mathematical models for the evolution of multigene families by unequal crossing over, *Nature* 265:304-310.

PERELSON, A. S., M. MIRMIRANI, and G. F. OSTER (1976): Optimal strategies in immunology. I: B-cell differentiation and proliferation, *J. Math. Biol.* 3:325-367.

______(1978): Optimal strategies in immunology. II: B memory cell production, *J. Math. Biol.* 5:213-256.

PERKINS, R. J. (1978): A strategy for species introductions into evolving model ecosystems, *Ecol. Model.* 5:1-15.

PETERS, R. H. (1976): Tautology in evolution and ecology, *Amer. Nat.* 110:1-12.

PETERSEN, R. (1975): The paradox of the plankton: an equilibrium hypothesis, *Amer. Nat.* 109:35-49.

PETRAITIS, P. S. (1979): Likelihood measures of niche bredth and overlap, *Ecology* 60:703-710.

PHILLIPS, O. M. (1973): The equilibrium and stability of simple marine biological systems. I: Primary nutrient consumers, *Amer. Nat.* 107:73-93.

______(1974): The equilibrium and stability of simple marine biological systems. II: Herbivores, *Arch. Hydrobiol.* 73: 310-333.

______(1978): The equilibrium and stability of simple marine biological systems III: Fluctuations and survival, *Amer. Nat.* 112:745-757.

PIELOU, E. C. (1969): *An Introduction to Mathematical Ecology*, Wiley-Interscience, New York.

______(1972): Niche width and niche overlap: a method for measuring them, *Ecology* 53:687-692.

______(1974): Competition on an environmental gradient, in P. van den Driessche, ed., *Mathematical Problems in Biomathematics*, Springer-Verlag Lecture Notes in Biomathematics, Vol. 2, New York, pp. 184-204.

PIMBLEY, G. H., Jr. (1974a): Periodic solutions of predator-prey equations simulating an immune response, I, *Math. Biosci.* 20: 27-51.

______(1974b): Periodic solutions of predator-prey equations simulating an immune response, II, *Math. Biosci.* 21:251-277.

______(1974c): Periodic solutions of third order predator-prey equations simulating an immune response, *Arch. Rat. Mech. Anal.* 55:93-123.

______(1977): Bifurcation behavior of periodic solutions for an immune response problem, *Arch. Rat. Mech. Anal.* 64:169-192.

PIMENTEL, D., S. A. LEVIN, and D. OLSON (1978): Coevolution and the stability of exploiter-victim systems, *Amer. Nat.* 112:119-125.

PODOLER, H., and Z. MENDEL (1979): Analysis of a host-parasite (Ceratitis-Muscidifurax) relationship under laboratory conditions, *Ecol. Ent.* 4:45-59.

POLYAKOV, I. Ya (1972): Ecological fundamentals of pest control in plants, *Sov. J. Ecol.* 3:305-314.

POMERANTZ, M. J., and M. E. GILPIN (1979): Community covariance and coexistence, *J. Theor. Biol.* 79:67-81.

POOLE, R. W., (1974): A discrete time stochastic model of a two prey, one predator species interaction, *Theor. Pop. Biol.* 5: 208-228.

______(1976a): Empirical multivariate autoregressive equation predictors of the fluctuations of interacting species, *Math. Biosci.* 28:81-97.

______(1976b): Stochastic difference equation predictors of population fluctuations, *Theor. Pop. Biol.* 9:25-45.

______(1977): Periodic, pseudoperiodic, and chaotic population fluctuation, *Ecology* 58:210-213.

POST, W. M., and C. C. TRAVIS (1979): Qualitative stability in models of ecological communities, *J. Theor. Biol.* 79:547-553.

POULSON, E. T. (1980): A model for population regulation with density-and frequency-dependent selection, *J. Math. Biol.* 9: 325-343.

POWER, H. M. (1978): Formal models for biological oscillators, derived by using Liapunov stability theory in a synthetic mode, *Proc. Roy. Irish Acad. A* 78:235-245.

POWERS, J. E., and R. T. LACKEY (1975): Interaction in ecosystems: a queuing approach to modeling, *Math. Biosci.* 25:81-90.

PRAJNESHU (1978): Two-species system in random environment, *Austr. J. Stat.* 20:275-281.

______(1979): Stability of two interacting species subject to stochastic parameter variation, *Commun. Stat. - Theor. Methods* A8:57-69.

PRESTON, F. W. (1969): Diversity and stability in the biological world, in *Diversity and Stability in Ecological Systems,* Brookhaven Symposia in Biology, Vol. 22, pp. 1-12.

PRIKHOD'HO, T. I. (1976): Mathematical modeling of populations of planktonic crustaceans, *Sov. J. Ecol.* 7:1-8.

PYKH, Iu. A. (1977): Stability of solutions of Lotka-Volterra differential equations, *Appl. Math. Mech.* 41:253-261.

RAFF, M. C. (1976): Cell-surface immunology, *Sci. Amer.* 234 (5): 30-39.

RAPOPORT, A. (1956): Some game-theoretical aspects of parasitism and symbiosis, *Bull. Math. Biophys.* 18:15-30.

RAPPORT, D. J., and J. E. TURNER (1975): Predator-prey interactions in natural communities, *J. Theor. Biol.* 51:169-180.

RATHCKE, B. J. (1976): Competition and coexistence within a guild of herbivorous insects, *Ecology* 57:76-87.

RAVEH, A., and U. RITTE (1976): Frequency dependence and stability, *Math. Biosci.* 30:371-374.

REAL, L. A. (1977): The kinetics of functional response, *Amer. Nat.* 111:289-300.

______(1979): Ecological determinants of functional response, *Ecology* 60:481-485.

REDDINGUS, J. (1963): A mathematical note on a model of a consumer-food relation in which the food is continually replaced, *Acta Biotheor.* 16:183-198.

REED, W. J. (1978): The steady state of a stochastic harvesting model, *Math. Biosci.* 41:273-307.

REJAMNEK, M., and J. JENIK (1975): Niche, habitat, and related ecological concepts, *Acta Biotheor.* 23:100-107.

RESCIGNO, A. (1968): The struggle for life. II: Three competitors, *Bull. Math. Biophys.* 30:291-297.

______(1977a): The struggle for life. IV: Two predators sharing a prey, *Bull. Math. Biol.* 39:179-185.

______(1977b): The struggle for life. V: One species living in a limited environment, *Bull. Math. Biol.* 39:479-485.

RESCIGNO, A., and C. DeLISI (1977): Immune surveillance and neoplasia. II: A two-stage mathematical model, *Bull. Math. Biol.* 39:487-497.

RESCIGNO, A., and K. G. JONES (1972): The struggle for life. III: A predator-prey chain, *Bull. Math. Biophys.* 34:521-532.

RESCIGNO, A., and I. W. RICHARDSON (1965): On the competitive exclusion principle, *Bull. Math. Biophys.* 27:85-89.

______(1967): The struggle for life. I: Two species, *Bull. Math. Biophys.* 29:377-388.

RHODE, K. (1979): A critical evaluation of intrinsic and extrinsic factors responsible for niche restriction in parasites, *Amer. Nat.* 114:648-671.

RICKLEFS, R. E. (1972): Dominance and the niche in bird communities, *Amer. Nat.* 106:538-545.

RICHTER, P. H. (1975): A network theory of the immune system, *Eur. J. Immunol.* 5:350-354.

RIDLER-ROWE, C. J. (1978): On competition between two species, *J. Appl. Prob.* 15:457-465.

RIEBESELL, J. F. (1974): Paradox of enrichment in competitive systems, *Ecology* 55:183-187.

ROCKLIN, S., and G. OSTER (1976): Competition between phenotype, *J. Math. Biol.* 3:225-261.

ROGERS, D. (1972): Random search and insect population models, *J. Anim. Ecol.* 41:369-383.

ROGERS, D. J., and M. P. HASSELL (1974): General models for insect parasite and predator searching behaviour: interference, *J. Anim. Ecol.* 43:239-253.

ROGERS, D. J., and S. HUBBARD (1974): How the behavior of parasites and predators promotes population stability, in M. B. Usher and M. H. Williamson, eds., *Ecological Stability*, Chapman and Hall, London, pp. 99-119.

RORRES, C. (1976): Stability of an age specific population with density dependent fertility, *Theor. Pop. Biol.* 10:26-46.

ROSEN, G. (1976): Construction of analytically solvable models for interacting species, *Bull. Math. Biol.* 38:193-197.

______(1977a): Effects of diffusion on the stability of the equilibrium in multi-species ecological systems, *Bull. Math. Biol.* 39:373-383.

______(1977b): On the persistence of ecological systems, *J. Theor. Biol.* 65:795-799.

ROSENBLAT, S. (1980): Population models in a periodically fluctuating environment, *J. Math. Biol.* 9:23-36.

ROSENZWEIG, M. L. (1969): Why the prey curve has a hump, *Amer. Nat.* 103:81-87.

______(1971): Paradox of enrichment: destabilization of exploitation ecosystems in ecological time, *Science* 171:385-387.

______(1972a): Reply to McAllister et al: "Stability of enriched aquatic ecosystems," *Science* 175:564-565.

______(1972b): Reply to Gilpin: "Enriched predator-prey systems: theoretical stability," *Science* 177:904.

______(1973a): Evolution of the predator isocline, *Evolution* 27: 84-94.

______(1973b): Exploitation in three trophic levels, *Amer. Nat.* 107:275-294.

ROSENZWEIG, M. L., and R. H. MACARTHUR (1963): Graphical representation and stability conditions of predator-prey interactions, *Amer. Nat.* 47:209-223.

ROSENZWEIG, M. L., and W. M. SCHAFFER (1978): Homage to the red queen II: Coevolutionary response to enrichment of exploitation ecosystems, *Theor. Pop. Biol.* 14:158-163.

ROSS, G. G. (1972): A difference-differential model in population dynamics, *J. Theor. Biol.* 34:477-492.

______(1973a): A model for the competitive growth of two diatoms, *J. Theor. Biol.* 42:307-331.

______(1973b): A population model for limited food competition, *J. Theor. Biol.* 42:333-347.

ROSS, G. G., and N. A. SLADE (1977): A predator-prey viewpoint of a single species population, *J. Theor. Biol.* 77:513-522.

ROSSLER, O. E., and K. WEGMANN (1978): Chaos in the Zhabotinskii reaction, *Nature* 271:80-90.

ROTENBERG, M. (1975): Equilibrium and stability in populations whose interactions are age-specific, *J. Theor. Biol.* 54:207-224.

ROUGHGARDEN, J. (1974a): Reply to Thoday, *Amer. Nat.* 108:143.

______(1974b): Species packing and the competition function with illustrations from coral reef fish, *Theor. Pop. Biol.* 5:163-186.

______(1974c): The fundamental and realized niche of a solitary population, *Amer. Nat.* 108:232-235.

______(1974d): Niche width: biogeographic patterns among Anolis lizard populations, *Amer. Nat.* 108:429-442.

______(1974e): Population dynamics in a spatially varying environment: how population size "tracks" spatial variation in carrying capacity, *Amer. Nat.* 108:649-664.

______(1975a): Population dynamics in a stochastic environment: spectral theory for the linearized N-species Lotka-Volterra competition equations, *Theor. Pop. Biol.* 7:1-12.

______(1975b): Species packing and predation pressure, *Ecology* 56:489-492.

______(1975c): A simple model for population dynamics in stochastic enviroments, *Amer. Nat.* 109:713-736.

______(1975d): Evolution of marine symbiosis - a simple cost-benefit model, *Ecology* 56:1201-1208.

______(1976): Resource partitioning among competing species - a coevolutionary approach, *Theor. Pop. Biol.* 9:388-424.

______(1977): Coevolution in ecological systems: results from "loop analysis" for purely density-dependent coevolution, in F. B. Christiansen and T. M. Fenchel, eds., *Measuring Selection in Natural Populations*, Springer-Verlag Lecture Notes in Biomathematics, Vol. 19, New York, pp. 499-517.

ROUTLEDGE, R. D. (1977): On Whittaker's components of diversity, *Ecology*, 58:1120-1127.

______(1979a): Niche metrics and diversity components, *Oecologia (Berl.)* 43:121-124.

______(1979b): Diversity indices: which ones are admissible? *J. Theor. Biol.* 76:503-515.

______(1980): The form of species - abundance distributions, *J. Theor. Biol.* 82:547-558.

ROYAMA, T. (1971): A comparative study of models of predation and parasitism, *Res. Pop. Ecol.*, Sup. I, pp. 1-91.

______(1977): Population persistence and density dependence, *Ecol. Monogr.* 47:1-35.

RUDD, W. G. (1975): Population modeling for pest management studies, *Math. Biosci.* 26:282-302.

RUDIN, W. (1953): *Principles of Mathematical Analysis*, McGraw-Hill, New York.

RUTLEDGE, R. W., B. L. BASORE, and R. J. MULHOLLAND (1976): Ecological stability: an information theory viewpoint, *J. Theor. Biol.* 57:355-371.

SABATH, M. D. (1974): Niche breadth and genetic variability in sympatric natural populations of drosophilid flies, *Amer. Nat.* 108:533-540.

SABATH, M. D., and J. M JONES (1973): Measurement of niche breadth and overlap: the Colwell-Futuyma method, *Ecology* 54:1143-1147.

SALT, G. W. (1966): An examination of logarithmic regression as a measure of population density response, *Ecology* 47:1035-1039.

______(1967): Predation in an experimental protozoan population *(Woodruffia-Paramecium)*, *Ecol. Monog.* 37:113-144.

______(1974): Predator and prey densities as controls of the rate of capture by the predator *Didinium nasutum*, *Ecology* 55:434-439.

SALT, G. W., and D. E. WILLARD (1971): The hunting behavior and success of Forster's tern, *Ecology* 52:989-998.

SAMUELSON, P. A. (1967): A universal cycle? *Oper. Res.* 3:307-320.

______(1971): Generalized predator-prey oscillations in ecological and economic equilibrium, *Proc. Nat. Acad. Sci. USA* 68:980-983.

______(1976): Time symmetry and asymmetry in population and determination dynamic systems, *Theor. Pop. Biol.* 9:82-122.

SANCHEZ, D. A. (1977): Populations and harvesting, *SIAM Rev.* 19:551-553.

______(1978): Linear age-dependent population growth with harvesting, *Bul. Math. Biol.* 40:377-385.

SAUNDERS, P. T. (1976): Oscillations in tropical ecosystems, *Nature* 261:525.

SAUNDERS, P. T., and M. J. BAZIN (1975): On the stability of food chains, *J. Theor. Biol.* 52:121-142.

SAUNDERS, P. T., J. H. LAWTON, and S. L. PIMM (1978): Population dynamics and the length of food chains: Reply, *Nature* 272:189-190.

SCHAFFER, W. M. (1977): Evolution, population dynamics and stability: a comment, *Theor. Pop. Biol.* 11:326-329.

SCHAFFER, W. M., and M. L. ROSENZWEIG (1977): Selection for optimal life histories. II: Multiple equilibria and the evolution of alternate reproductive strategies, *Ecology* 58:60-72.

______(1978): Homage to the red queen I: Coevolution of predators and their victims, *Theor. Pop. Biol.* 14:135-157.

SCHNEIDER, D. (1978): Equalization of prey numbers by migratory shorebirds, *Nature* 271:353-354.

SCHNUTE, J., and P. VAN DEN DRIESSCHE (1975): Two biological applications of Dulac's criterion, *Appl. Math. Notes* 1:75-81.

SCHOENER, T. W. (1973): Population growth regulated by intraspecific competition for energy or time: some simple representations, *Theor. Pop. Biol.* 4:56-84.

______(1974a): Competition and the form of habitat shift, *Theor. Pop. Biol.* 6:265-307.

______(1974b): Some methods for calculating competition coefficients from resource-utilization spectra, *Amer. Nat.* 108:332-340.

______(1976): Alternatives to Lotka-Volterra competition: models of intermediate complexity, *Theor. Pop. Biol.* 10:309-333.

______(1978): Effects of density-restricted food encounter on some single-level competition models, *Theor. Pop. Biol.* 13:365-381.

SCHOENER, T. W., R. B. HUEY, and E. R. PIANKA (1979): A biogeographic extension of the compression hypothesis: competitors in narrow sympatry, *Amer. Nat.* 113:295-298.

SCHUSTER, P., K. SIGMUND, and R. WOLFF (1979): Dynamical systems under constant organization. III: Cooperative and competitive behavior of hypercycles, *J. Differential Eqns.* 32:357-368.

SCUDO, F. M. (1971): Vito Volterra and theoretical ecology, *Theor. Pop. Biol.* 2:1-23.

SCUDO, F. M., and J. R. ZIEGLER (1976): Vladimir Alexandrovich Kostitzin and theoretical ecology, *Theor. Pop. Biol.* 10:395-412.

SEGAL, L. A., and S. A. LEVIN (1976): Application of nonlinear stability theory to the study of the effects of diffusion on predator-prey interactions, in R. A. Piccirelli, ed., *Topics in Statistical Mechanics and Biophysics: A Memorial to Julius L. Jackson, Proc. AIP Conf.* 27:123-152.

SEIFERT, R. P., and F. H. SEIFERT (1976): A community matrix analysis of *Heliconia* insect communities, *Amer. Nat.* 110:461-483.

SEIP, K. L. (1980): A computational model for growth and harvesting of the marine alga Ascophyllum nodosum, *Ecol. Model.* 8:189-199.

SHAPIRO, A. P. (1974a): A model of competing species, *Dokl. Akad. Nauk SSSR* 215:111-112 (translation).

______ (1974b): A discrete model of the competition of two populations, *Dokl. Bio. Sci.* 28:406-408.

SHEPPARD, D. H. (1971): Competition between two chipmunk species *(Eutamias)*, *Ecology* 52:320-329.

SHIGESADA, N. (1980): Spatial distribution of dispersing animals, *J. Math. Biol.* 9:85-96.

SHIGESADA, N., and E. TERAMOTO (1978): A consideration on the theory of environmental density, *Jap. J. Ecol.* 28:1-8 (Japanese).

SHIGESADA, N., K. KAWASAKI, and E. TERAMATO (1979): Spatial segregation of interacting species, *J. Theor. Biol.* 79:83-99.

SHILEPSKY, C. C. (1974): The asymptotic behavior of an integral equation with an application to Volterra's population equation, *J. Math. Anal. Appl.* 48:764-779.

SHIRAKIHARA, K., and S. TANAKA (1978): Two fish species competition model with nonlinear interactions and equilibrium catches, *Res. Pop. Ecol.* 20:123-140.

SHOCK, N. W., and M. F. MORALES (1942): A fundamental form for the differential equation of colonial and organism growth, *Bull. Math. Biophys.* 4:63-71.

SHORROCKS, B., W. ATKINSON, and P. CHARLESWORTH, Competition on a divided and ephemeral resource, *J. Anim. Ecol.* 48:899-908.

SILJAK, D. D. (1975): When is a complex ecosystem stable? *Math. Biosci.* 25:25-50.

SILVERT, W. (1978): Anomalous enhancement of mean population levels by harvesting, *Math. Biosci.* 42:253-256.

SIMBERLOFF, D. S., and L. G. ABELE (1976): Island biogeography theory and conservation practice, *Science* 191:285-286.

SKELLAM, J. G., M. V. BRIAN, and J. R. PROCTOR (1960): The simultaneous growth of interacting systems, *Acta Biotheor.* 13:131-144.

SLATKIN, M. (1974): Competition and regional coexistence, *Ecology* 55:128-134.

______(1978a): The dynamics of a population in a markovian environment, *Ecology* 59:249-256.

______(1978b): On the equilibrium of fitness by natural selection, *Amer. Nat.* 112:845-859.

SLATKEN, M., and R. LANDE (1976): Niche width in a fluctuating environment-density independent model, *Amer. Nat.* 110:31-55.

SLATKIN, M., and D. S. WILSON (1979): Coevolution in structured demes, *Proc. Nat. Acad. Sci. USA* 76:2084-2087.

SLOBODKIN, L. B. (1961): Preliminary ideas for a predictive theory of ecology, *Amer. Nat.* 95:147-153.

______(1974): Prudent predation does not require group selection, *Amer. Nat.* 108:665-678.

SLOBODKIN, L. B., F. E. SMITH, and N. G. HAIRSTON (1967): Regulation in terrestrial ecosystems and the implied balance of nature, *Amer. Nat.* 101:109-124.

SMALE, S. (1976): On the differential equations of species in competition, *J. Math. Biol.* 3:5-7.

SMALE, S., and R. F. WILLIAMS (1976): The qualitative analysis of a difference equation of population growth, *J. Math. Biol.* 3:1-4.

SMITH, C. E., and H. C. TUCKWELL (1974): Some stochastic growth processes, in P. van den Driessche, ed., *Mathematical Problems in Biology*, Springer-Verlag Lecture Notes in Biomathematics, Vol. 2, New York, pp. 211-225.

SMITH, H. L. (preprint): Season-caused coexistence in a competition model.

SMITH, O. L., H. H. SHUGART, R. V. O'NEILL, R. S. BOOTH, and D. C. McNAUGHT (1975): Resource competition and an analytical model of zooplankton feeding on phytoplankton, *Amer. Nat.* 109:571-591.

SMITH, R. H., and R. MEAD (1974): Age structure and stability in models of prey-predator systems, *Theor. Pop. Biol.* 6:308-322.

______(1979): On predicting extinction in simple population models I: Stochastic linearization, *J. Theor. Biol.* 80:189-203.

SMITH-GILL, S. J., and D. E. GILL (1978): Curvelinearities in the competition equations: an experiment with hanid tadpoles, *Amer. Nat.* 112:557-570.

SMOUSE, P. E. (1976): The implications of density-dependent population growth for frequency- and density-dependent selection, *Amer. Nat.* 110:849-860.

SO, J. W. H. (1979): A note on the global stability and bifurcation phenomenon of a Lotka-Volterra food chain, *J. Theor. Biol.* 80:185-187.

SOLOMON, M. E. (1968): Logarithmic regression as a measure of population density response: comment on a report by G. W. Salt, *Ecology* 49:357-358.

SOLOMON, M. E., and D. M. GLEN (1979): Prey density and rates of predation by tits (Parus spp.) on larvae of codling moth (Cydia pomonella) under bark, *J. Appl. Ecol.* 16:49-59.

SOWUNMI, C. O. A. (1976): Female dominant age-dependent deterministic population dynamics, *J. Math. Biol.* 3:9-17.

STEELE, J. (1974): Stability of plankton ecosystems, in M. B. Usher and M. H. Williamson, eds., *Ecological Stability*, Chapman and Hall, London, pp. 179-191.

______(1976): Application of theoretical models in ecology, *J. Theor. Biol.* 63:443-451.

STEPHANOPOULOS, G., R. ARIS, and A. G. FREDRICKSON (1979): A stochastic analysis of the growth of competing microbial populations in a continuous biochemical reactor, *Math. Biosci.* 45:99-135.

STEWART, F. M., and B. R. LEVIN (1973): Partitioning of resources and the outcome of interspecific competition: a model and some general considerations, *Amer. Nat.* 107:171-198.

STIRZAKER, D. (1975): On a population model, *Math. Biosci.* 23:329-336.

STREBEL, D. E., and N. S. GOEL (1973): On the isocline methods for analyzing prey-predator interactions, *J. Theor. Biol.* 39:211-239.

STREIFER, W., and C. A. ISTOCK (1973): A critical variable formulation of population dynamics, *Ecology* 54:392-398.

STRICKFADEN, W. B., and B. A. LAWRENCE (1975): Solvable limit cycle in a Volterra-type model of interacting populations, *Math. Biosci.* 23:273-279.

STROBECK, C. N. (1973): N species competition, *Ecology* 54:650-654.

STUBBS, M. (1977): Density dependence in the life-cycle of animals and its importance in K- and r- strategies, *J. Anim. Ecol.* 46:677-688.

SUDO, R., K. KOBAYASHI, and S. AIBA (1975): Some experiments and analysis of a predator-prey model: interaction between *Colpidium campylum* and *Alcaligenes faecalis* in continuous and mixed culture, *Biotech. Bioeng.* 17:167-184.

SWICK, K. E. (1976): A model of single species population growth, *SIAM J. Math. Anal.* 7:565-576.

______(1977): A nonlinear age-dependent model of single species populations dynamics, *SIAM J. Appl. Math.* 32:484-498.

SYKES, R. M. (1974): Competition for nutrients among the plankton, *Amer. Nat.* 108:578-579.

TAKAFUJI, A., and D. A. CHANT (1976): Comparative studies of two species predacious phytoseiid mites (Acarinai Phytoseiidae), with special reference to their responses to the density of their prey, *Res. Pop. Ecol.* 17:255-310.

TAKEUCHI, Y., N. ADACHI, and H. TOKUMARU (1978a): The stability of generalized Volterra equations, *J. Math. Anal Appl.* 68:453-473.

______(1978b): Global stability of ecosystems of the generalized Volterra type, *Math. Biosci.* 42:119-136.

TAMARIN, R. H. (1978): Dispersal, population regulation, and K-selection in field mice, *Amer. Nat.* 112:545-555.

TANG, M. M., and P. C. FIFE (1980): Propagating fronts for competing species with diffusion, *Arch. Rat. Mech. Anal.* 73: 69-77.

TANNER, J. T. (1966): Effects of population density on growth rates of animal populations, *Ecology* 47:733-745.

______(1975): The stability and the intrinsic growth rates of prey and predator populations, *Ecology* 56:855-867.

TANSKY, M. (1978a): Stability of multispecies prey-predator system, *Mem. Coll. Sci. Univ. Kyoto* 7:87-94.

______(1978b): Switching effect in prey-predator system, *J. Theor. Biol.* 70:263-271.

TAYLOR, C. E., and R. R. SOKAL (1976): Oscillations in housefly population sizes due to time lags, *Ecology* 57:1060-1061.

TAYLOR, F. (1979): On the applicability of current population models to the growth of insect populations, *Rocky Mountain J. Math.* 9:149-151.

TAYLOR, N. W. (1968): A mathematical model for two Tribolium populations in competition, *Ecology* 49:843-848.

TAYLOR, R. J. (1974): Role of learning in insect parasitism, *Ecol. Monogr.* 44:89-104.

______(1976): Value of clumping in prey and the evolutionary response of ambush predators, *Amer. Nat.* 110:13-29.

______(1979): The value of clumping to prey when detectability increases with group size, *Amer. Nat.* 113:299-301.

TELFER, E. S. (1971): Changes in carrying capacity of deer range in western Nova Scotia, *Can. Field Nat.* 85:231-234.

TERAMOTO, E., K. KAWASAKI, and N. SHIGESADA (1979): Switching effect of predation on competitive prey species, *J. Theor. Biol.* 79:303-315.

THODAY, J. M. (1974): Evolution of niche width, *Amer. Nat.* 108: 142-143.

THORNTON, K. W., and R. J. MULHOLLAND (1974): Lagrange stability and ecological systems, *J. Theor. Biol.* 45:473-485.

THOMPSON, D. J. (1978): Towards a realistic predator-prey model: the effect of temperature on the functional response and life history of larvae of the damselfly, Ischnura elegans, *J. Anim. Ecol.* 47:757-767.

THOMPSON, M. (1978): Asymptotic growth and stability in populations with time dependent vital rates, *Math. Biosci.* 42:267-278.

TIMIN, M. E., and B. D. COLLIER (1971): A model incorporating energy utilization for the dynamics of single species populations, *Theor. Pop. Biol.* 2:237-251.

TINNIN, R. O. (1972): Interference or competition? *Amer. Nat.* 106:672-675.

TITMAN, D. (1976): Ecological competition between algae: experimental confirmation of resource-based competition theory, *Science* 192:463-465.

TIWARI, J. L., and J. E. HOBBIE (1976): Random differential equations as models of ecosystems: Monte Carlo simulation approach, *Math. Biosci.* 28:25-44.

TOGNETTI, K. (1975): The two stage stochastic population model, *Math. Biosci.* 25:195-204.

TRAVIS, C. C., and W. M. POST, III (1979): Dynamics and comparative statics of mutualistic communities, *J. Theor. Biol.* 78:553-571.

TREGONNING, K., and A. ROBERTS (1978): Ecosystem-like behavior of a random-interaction model I, *Bull. Math. Biol.* 40:513-524.

TRUBATCH, S. L., and A. FRANCO (1974): Canonical procedures for population dynamics, *J. Theor. Biol.* 48:299-324.

TULJAPULKAR, S. D., and J. S. SEMURA (1979): Liapunov functions: geometry and stability, *J. Math. Biol.* 8:25-32.

TURELLI, M. (1978a): A reexamination of stability in randomly varying versus deterministic environments with comments on the stochastic theory of limiting similarity, *Theor. Pop. Biol.* 13:244-267.

______(1978b): Does environmental variability limit niche overlap? *Proc. Nat. Acad. Sci. USA* 75:5085-5089.

TURNER, J. E., and D. J. RAPPORT (1974): An economic model of population growth and competition in natural communities, in P. van den Driessche, ed., *Mathematical Problems in Biology*, Springer-Verlag Lecture Notes in Biomathematics, Vol. 2, New York, pp. 236-240.

TURNER, M. E., Jr., B. A. BLUMENSTEIN, and J. L. SEBAUGH (1969): A generalization of the logistic law of growth, *Biometrics* 25: 577-580.

TURNER, M. E. Jr., E. L. BRADLEY, Jr., K. A. KIRK, and K. M. PRUITT (1976): A theory of growth, *Math. Biosci.* 29:367-373.

TUTUBALIN, V. M. (1973): On dynamic systems with random perturbations, *SIAM Theor. Prob.* 18:678-693.

ULANOWICZ, R. E. (1972): Mass and energy flow in closed ecosystems, *J. Theor. Biol.* 34:239-252.

ULANOWICZ, R. E., D. A. FLEMER, D. R. HINLE, and R. T. HUFF (1978): The empirical modeling of an ecosystem, *Ecol. Model.* 4:29-40.

UTIDA, S. (1953): Interspecific competition between two species of bean weevil, *Ecology* 34:301-307.

______(1957): Cyclic fluctuations of population density intrinsic to the host-parasite system, *Ecology* 38:442-449.

UTZ, W. R. (1961): The equations of population growth, *Bull. Math. Biophys.* 23:261-262.

UTZ, W. R., and P. E. WALTMAN (1963): Periodicity and boundedness of solutions of generalized differential equations of growth, *Bull. Math. Biophys.* 25:75-93.

VANCE, R. R. (1978): Predation and resource partitioning in one predator-two prey model communities, *Amer. Nat.* 112:797-813.

VAN DEN DRIESSCHE, P. (1974): Stability of linear population models, *J. Theor. Biol.* 48:473-476.

VAN DEN ENDE, P. (1973): Predator-prey interactions in continuous culture, *Science* 181:562-564.

VANDERMEER, J. H. (1969): The competitive structure of communities: an experimental approach with protozoa, *Ecology* 50:362-371.

______(1970): The community matrix and the number of species in a community, *Amer. Nat.* 104:73-83.

______(1973a): On the regional stabilization of locally unstable predator-prey relationships, *J. Theor. Biol.* 41:161-170.

______(1973b): Generalized models of two species interactions: a graphical analysis, *Ecology* 54:809-818.

______(1975): Interspecific competition: a new approach to the classical theory, *Science* 188:253-255.

VANDERMEER, J. H., and D. H. BOUCHER (1978): Varieties of mutualistic interaction in population models, *J. Theor. Ecol.* 74: 549-558.

VARMA, V. S. (1977): Exact solutions for a special prey-predator or competing system, *Bull. Math. Biol.* 39:619-622.

VAUGHAN, T. A., and R. M. HANSEN (1974): Experiments on interspecific competition between two species of pocket gophers, *Amer. Midl. Nat.* 45:444-452.

VEILLEUX, B. G. (1979): An analysis between the predatory interaction between Paramecium and Didinium, *J. Anim. Ecol.* 48: 787-803.

VERHULST, F. F. (1838): Notice sur la loi que la population suit dans son acroissement, *Corr. Math. Phys.* 10:113.

VILLARREAL, E., Z. AKCASU, and R. P. CANALE (1976): A theory of interacting microbial populations: multigroup approach, *J. Theor. Biol.* 58:285-317.

VINCENT, T. L., and L. R. ANDERSON (1979): Return time and vulnerability for a food chain model, *Theor. Pop. Biol.* 15: 217-231.

VOLTERRA, V. (1927): Variazioni e fluttuazioni del numero d'individui in specie animali conviventi, *Mem. R. Com. Talassogr. Ital.* 131:1-142.

______(1931): Lecons sur la theorie mathematique de la lutte pour la vie, *Gauthier-Villars*, Paris.

______(1937): Principes de biologie mathematique, *Acta Biotheor.* 3:1-36.

WALDON, M. G. (1975): Competition models, *Amer. Nat.* 109:487-489.

WALKER, J., A. VAN NYPELSEER, and W. E. LANGOIS (1976): Numerical integration of a stochastic model for the Volterra-Lotka reaction, *Bull. Math. Biol.* 38:535-546.

WALLACE, M. M. H., and M. C. WALTERS (1974): The introduction of *Bdellodes lapidaria* (Acari: Bdellidae) from Australia into South Africa for the biological control of *Sminthurus viridis* (Collembola), *Aust. J. Zool.* 22:505-517.

WALSH, G. R. (1978): Optimal control of pests in the presence of predators, *Bull. Math. Biol.* 40:319-333.

WALTER, C. (1974): The global asymptotic stability of prey-predator systems with second-order dissipation, *Bull. Math. Biol.* 36: 215-217.

WALTHER, H. -O. (1976): On a transcendental equation in the stability analysis of a population growth model, *J. Math. Biol.* 3:187-195.

WALTMAN, P. E. (1964): The equations of growth, *Bull. Math. Biophys.* 26:39-43.

WALTMAN, P. E., and E. BUTZ (1977): A threshold model of antigen-antibody dynamics, *J. Theor. Biol.* 65:499-512.

WALTMAN, P. E., S. P. HUBBELL, and S. B. HSU (preprint): Theoretical and experimental investigations of microbial competition in continuous culture.

WANG, F. J. S. (1975): Limit theorems for age and density dependent stochastic population models, *J. Math. Biol.* 2:373-400.

WANGERSKY, P. J. (1978): Lotka-Volterra population models, *Ann. Rev. Ecol. Syst.* 9:189-218.

WANGERSKY, P. J., and W. J. CUNNINGHAM (1956): On time lags in equations of growth, *Proc. Nat. Acad. Sci. USA* 42:699-702.

______(1957a): Time lag in population modes, *Cold Spring Harbor Symp. Qual. Biol.* 22:329-338.

______(1957b): Time lag in prey-predator population models, *Ecology* 38:136-139.

WATKINSON, A. R. (1980): Density-dependence in single-species populations of plants, *J. Theor. Biol.* 83:345-357.

WATT, K. E. F. (1959): A mathematical model for the effect of densities of attacked and attacking species on the number attacked, *Can. Ent.* 91:129-144.

______(1961): Mathematical models for use in insect pest control, *Can. Ent.* (suppl.) 19:1-62.

______(1964): Density dependence in population fluctuations, *Can. Ent.* 96:1147-1148.

______(1965): Community stability and the strategy of biological control, *Can. Ent.* 97:887-895.

WEBBER, M. I. (1974): Food web linkage complexity and stability in a model ecosystem, in M. G. Usher and M. H. Williamson, eds. *Ecological Stability*, Chapman and Hall, London, pp. 165-176.

WEINSTEIN, M. S. (1970): Hares, lynx, and trappers, *Amer. Nat.* 111:806-808.

WEISS, G. H. (1963): Comparison of a deterministic and a stochastic model for interaction between antagonistic species *Biometrics* 19:595-602.

______(1968): Equations for the age structure of growing populations, *Bull. Math. Biophys.* 30:427-435.

WESTERBERG, J. (1960): An analysis in population systems, *Acta Biotheor.* 13:145-160.

WHITE, B. S. (1977): The effects of a rapidly-fluctuating random environment on systems of interacting species, *SIAM J. Appl. Math.* 32:666-693.

WHITE, E. G., and C. B. HUFFAKER (1969a): Regulatory processes and population cyclicity in laboratory populations of *Anagasta kuhniella* (Zeller) (Lepidoptera: Phycitidae). I: Competition for food and predation, *Res. Pop. Ecol.* 11:57-83.

______(1969b): Regulatory processes and population cyclicity in laboratory populations of *Anagasta kuhniella* (Zeller) (Lepidoptera: Phycitidae). II: Parasitism, predation, competition and protective cover, *Res. Pop. Ecol.* 11:150-155.

WHITTAKER, R. H., and S. A. LEVIN (1977): The role of mosaic phenomena in natural communities, *Theor. Pop. Biol.* 12:117-139.

WHITTAKER, R. H., S. A. LEVIN, and R. B. ROOT (1973): Niche, habitat, and ecotype, *Amer. Nat.* 107:321-338.

WICKWIRE, K. (1977): Mathematical models for the control of pests and infectious diseases: a survey, *Theor. Pop. Biol.* 11:182-238.

WIEGERT, R. G. (1974): Competition: a theory based on realistic, general equations of population growth, *Science* 185:539-542.

WILBUR, H. M. (1972): Competition, predation, and the structure of the Ambystoma - Rana sylvatica community, *Ecology* 53:3-21.

WILCOX, D. L., and J. W. MacCLUER (1979): Coevolution in predator-prey systems: a saturation kinetic model, *Amer. Nat.* 113:163-183.

WILLIAMSON, M. H. (1974): The analysis of discrete time cycles, in M. B. Usher and M. H. Williamson, eds., *Ecological Stability*, Chapman and Hall, London, pp. 17-33.

WINKELHAKE, J. L. (1976): A dynamic model for molecular regulation of the humoral immune response, *J. Theor. Biol.* 60:37-49.

WITTEN, M. (1978): Fitness and survival in logistic models, *J. Theor. Biol.* 74:23-32.

WOLKIND, D. J. (1976): Exploitation in three trophic levels: an extension allowing intra-species carnivore interaction, *Amer. Nat.* 110:431-447.

WOLKIND, D. J., and J. A. LOGAN (1978): Temperature-dependent predator-prey mite ecosystem on apple tree foliage, *J. Math. Biol.* 6:265-283.

WÖRZ-BUSEKROS, A. (1978): Global stability in ecological systems with continuous time delay, *SIAM J. Appl. Math.* 35:123-134.

WU, L. S.-Y. (1977): The stability of ecosystems - a finite-time approach, *J. Theor. Biol.* 66:345-359.

WUENSCHER, J. E. (1969): Niche specification and competition modelling, *J. Theor. Biol.* 25:436-443.

YODZIS, P. (1976a): The effects of harvesting on competitive systems, *Bull. Math. Biol.* 38:97-109.

______(1976b): Species richness and stability of space-limited communities, *Nature* 264:540-541.

______(1977): Harvesting and limiting similarity, *Amer. Nat.*, 111: 833-843.

______(1978): *Competition for Space and the Structure of Ecological Communities*, Lecture Notes in Biomathematics, No. 25, Springer-Verlag, New York.

YORKE, J. A., and W. N. ANDERSON, Jr. (1973): Predator-prey patterns, *Proc. Nat. Acad. Sci. USA* 70:2069-2071.

ZARET, T. M. (1972): Predator-prey interactions in a tropical lacustrive ecosystem, *Ecology* 53:248-257.

ZARET, T. M., and A. S. RAND (1971): Competition in tropical stream fishes: support for the competitive exclusion principle, *Ecology* 52:336-342.

ZEIGLER, B. (1977): Persistence and patchiness of predator-prey systems induced by discrete event population exchange mechanisms, *J. Theor. Biol.* 67:687-713.

______(1978): On necessary and sufficient conditions for group selection efficacy, *Theor. Pop. Biol.* 13:356-364.

ZICARELLI, J. D. (1975): Mathematical analysis of a population model with several predators on a single prey, Ph. D. Thesis, University of Minnesota, Minneapolis.

ZWANIG, R. (1973): Generalized Verhulst laws for population growth, *Proc. Nat. Acad. Sci. USA* 70:3048-3051.

INDEX